Distributed .NET with Microsoft Orleans

Build robust and highly scalable distributed
applications without worrying about complex
programming patterns

Bhupesh Guptha Muthiyalu

Suneel Kumar Kunani

BIRMINGHAM—MUMBAI

Distributed .NET with Microsoft Orleans

Copyright © 2022 Packt Publishing

Group Product Manager: Richa Tripathi
Publishing Product Manager: Kushal Dave
Senior Editor: Keagan Carneiro
Content Development Editor: Adrija Mitra
Technical Editor: Joseph Aloocaran
Copy Editor: Safis Editing
Project Coordinator: Rashika Ba
Proofreader: Safis Editing
Indexer: Manju Arasan
Production Designer: Ponraj Dhandapani
Marketing Coordinator: Pooja Yadav

First published: May 2022

Production reference: 2240522

Published by Packt Publishing Ltd.
Livery Place
35 Livery Street
Birmingham
B3 2PB, UK.

978-1-80181-897-1

www.packt.com

To my parents, Mrs. Geetha Rani and Mr. Muthiyalu, for the inspiration.
To my wife, Aparna, and my children, Tarak and Yashvak, for their
continued support and love. To my sister Janani for encouragement
throughout the long process of writing this book.

– Bhupesh Guptha Muthiyalu

To my lovely sister Suhasini, as well as my wife, Sravani, and children,
Mahanya and Krithi, for their unconditional love and support.

– Suneel Kumar Kunani

Foreword

As businesses continue to shift computational workloads into the cloud, it has become increasingly vital for developers to understand and overcome the challenges of building cloud-native applications. Such applications must gracefully survive machine failures, networking blips, rolling upgrades, and other disruptions. They must scale elastically and efficiently utilize the resources that are available to them, resources that are frequently in flux. These challenges distinguish the cloud from traditional, so-called on-premises applications. Tackling these challenges can be a daunting task, typically requiring application developers to become intimately familiar with distributed systems theory and practice. In response to this, Microsoft Research embarked on a project to simplify cloud application development such that programmers who are not experts in distributed systems can productively create resilient, scalable, and efficient cloud applications. The result is Microsoft Orleans.

In 2015, Orleans was released as an open source project. Having been introduced to Orleans during my previous time working at Microsoft, I became an avid user and contributor to the project and eventually rejoined Microsoft to develop Orleans full-time. It has been my pleasure to help the project mature, lead the project through our recent transition to the .NET team at Microsoft, and work with the many teams using Orleans within Microsoft and outside of the company.

I first came to know *Suneel* and *Bhupesh* while they were building cloud services using Orleans at Microsoft and I was delighted to learn sometime later that they were planning to write a book to help others to successfully leverage Orleans.

Over the course of *Distributed .NET with Microsoft Orleans*, Suneel and Bhupesh will guide you through building, testing, deploying, and monitoring real-world cloud applications using Orleans and .NET. They will explore the many features of Orleans, recommended design patterns for cloud applications in general and for Orleans-based applications, as well as the details of Orleans's programming model and runtime.

With *Distributed .NET with Microsoft Orleans*, you'll learn the best practices for writing high-quality, reliable, scalable distributed applications with Microsoft Orleans. After you complete Suneel and Bhupesh's book, you'll have developed a foundation for understanding distributed applications and the challenges of building for the cloud, and will have developed a real-world cloud-based application and deployed it to Azure Kubernetes Service and Azure App Service. This knowledge and experience will serve you well as you go on to build future applications for the cloud using Microsoft Orleans and .NET.

Elevate your skills into the stratosphere and build resilient, scalable, and efficient cloud-native applications with Orleans.

Reuben Bond

Software Engineer at Microsoft

Contributors

About the authors

Bhupesh Guptha Muthiyalu is a Microsoft Certified Professional and works at the company as a principal software engineering manager. He has 17+ years of software development experience on the .NET technology stack. His current role involves designing systems that are resilient to the iterations and changes required by the needs of enterprise businesses, validating architectural innovations, delivering solutions with high quality, managing the end-to-end ownership of products, and building diverse teams with capabilities to fulfill customer objectives. He is passionate about creating reusable components and identifying opportunities to make a product better.

"I would like to thank Sergey Bykov and Reuben Bond for their perspectives and suggestions to make this book better. I would like to extend my thanks to the Packt team, the co-authors, and the technical reviewers for the great partnership and collaboration. I am thankful to my managers and peers who supported and inspired me to write this book."

Suneel Kumar Kunani is a passionate developer who strives to learn something new every day. With over 17 years of experience in .NET and Microsoft technologies, he works on architecting and building mission-critical, highly scalable, and secure solutions at Microsoft. He loves to teach and preach the best practices in building distributed cloud solutions.

"My heartfelt gratitude goes out to Sergey Bykov and Reuben Bond for their unwavering support. I'd want to express my gratitude to the Packt team and reviewers for making the process of authoring this book so pleasant. I am thankful to my managers and colleagues for their support and encouragement as I worked on this book."

About the reviewers

Russ Hammett is an application developer and architect who is always trying to discover new ways of doing things. He previously got started working with and blogging about Orleans while working on the **Automated Cryptographic Validation Protocol (ACVP)** project under a Huntington Ingalls Industries contract for **National Institute of Standards and Technology (NIST)**. While he has not used Orleans in the "traditional sense" (his own words), he hopes that by continuing to blog and read about Orleans, one day he will get there! He believes that having as many tools in your toolbox as you can helps to uncover new approaches to solving problems, so expand your horizons and go beyond your comfort zone!

Separate from his career, Russ likes to spend time with his family, cook, and play video games/blog under the handle Kritner.

Jyothsna Shenoy is a principal software engineer with over 10 years' experience in various technologies, including .NET Framework, JavaScript, Azure, and various web development frameworks. She has architected, designed, and developed a wide variety of applications, from banking and **Business Process Management (BPM)** enterprise tools to consumer-facing mobile applications. In her spare time, she is an avid science-fiction reader.

Table of Contents

2

Cloud Architecture and Patterns for Distributed Applications

Section 2 - Working with Microsoft Orleans

3

Introduction to Microsoft Orleans

4
Understanding Grains and Silos

5
Persistence in Grains

6
Scheduling and Notifying in Orleans

Section 4 - Hosting and Deploying Orleans Applications to Azure

10

Deploying an Orleans Application in Azure Kubernetes

11

Deploying an Orleans Application to Azure App Service

Index

Other Books You May Enjoy

Preface

Building distributed applications in the modern era can be a tedious task, as customers expect high availability, high performance, and improved resilience. With the help of this book, you'll discover how you can harness the power of Microsoft Orleans to build impressive distributed applications.

Distributed .NET with Microsoft Orleans will demonstrate how to leverage Orleans to build highly scalable distributed applications step by step in the least possible time and with minimal effort. You'll explore Microsoft Orleans' key concepts such as the Orleans programming model, runtime, virtual actor, hosting, and deployment. As you advance, you'll become well-versed with important Orleans concepts such as grains, silos, timers, and persistence. Throughout the book, you'll create an application that will help you explore and learn Microsoft Orleans.

Who this book is for

This is intended for .NET developers and software architects who want a simple guide to creating distributed applications without having to worry about complex programming patterns. This book will also be useful for intermediate web developers who want to build highly scalable distributed applications. It is beneficial to have a basic understanding of .NET with C# and Azure.

What this book covers

Chapter 1, An Introduction to Distributed Applications, covers the fundamentals of distributed applications and serves as a fast tour of the distributed landscape.

Chapter 2, Cloud Architecture and Patterns for Distributed Applications, goes through the various cloud architectural patterns that can be used to create distributed applications and explains how they work.

Chapter 3, Introduction to Microsoft Orleans, introduces you to Microsoft Orleans and how it was created using Microsoft Research's Virtual Actor model concept.

Chapter 4, Understanding Grains and Silos, introduces grains, the primary building block of Orleans. You will learn about grain identification and how code generation works in Microsoft Orleans. Silo, and its lifespan and configuration, will also be covered.

Chapter 5, Persistence in Grains, will teach you about grain state persistence. Using an in-memory database, Azure Cosmos DB, and Azure Table storage, we will learn about distinct grain state storage components. We'll also learn how to have several states for grain. We'll show you how to build your own custom state provider. Then, we will learn how to directly interact with databases using patterns like **repository**.

Chapter 6, Scheduling and Notifying in Orleans, explores the role of different scheduling methods available in Orleans, timers and reminders. We'll look at the distinctions between them and how they're used.

Chapter 7, Engineering Fundamentals in Orleans, covers engineering principles such as unit testing, telemetry, and monitoring. For telemetry, we'll be using Azure Application Insights.

Chapter 8, Advanced Concepts in Orleans, explores some advanced concepts in Orleans such as streaming, grain interface versioning, and heterogeneous silos.

Chapter 9, Design Patterns in Orleans, delves into the most commonly used design patterns utilized in Orleans. Observer, Dispatcher, Smart Cache, and Reduce are the patterns we'll study.

Chapter 10, Deploying an Orleans Application in Azure Kubernetes, walks through how to host and deploy our Orleans application in Azure Kubernetes.

Chapter 11, Deploying an Orleans Application to Azure App Service, shows you how to use Azure App Service to host an Orleans application.

To get the most out of this book

You need to have the .NET 6 SDK installed on your system; all the code samples are tested using Visual Studio 2022 on the Windows OS. It is recommended to have an active Azure subscription.

Software/hardware covered in the book:

- C#
- .NET 6
- Visual Studio 2022
- Microsoft Azure

Operating system requirement: Windows

Download the example code files

You can download the example code files for this book from GitHub at `https://github.com/PacktPublishing/Distributed-.NET-with-Microsoft-Orleans`. If there's an update to the code, it will be updated in the GitHub repository.

We also have other code bundles from our rich catalog of books and videos available at `https://github.com/PacktPublishing/`. Check them out!

Download the color images

We also provide a PDF file that has color images of the screenshots and diagrams used in this book. You can download it here: `https://static.packt-cdn.com/downloads/9781801818971_ColorImages.pdf`.

Conventions used

There are a number of text conventions used throughout this book.

`Code in text`: Indicates code words in text, database table names, folder names, filenames, file extensions, pathnames, dummy URLs, user input, and Twitter handles. Here is an example: "The `reminderName` string is a string that uniquely identifies the reminder within the scope of the contextual grain."

A block of code is set as follows:

```
var silo = new SiloHostBuilder()
    [...]
    .UseInMemoryReminderService()
    [...]
```

When we wish to draw your attention to a particular part of a code block, the relevant lines or items are set in bold:

```
public interface IHotelGrain : IGrainWithStringKey
    {
        <<Code removed for brevity>>
        public Task Subscribe(IObserver observer);
        public Task UnSubscribe(IObserver observer);
```

Any command-line input or output is written as follows:

```
az monitor app-insights component create --app ai-distel-prod
--location westus  --resource-group rg-distel-prod
```

Bold: Indicates a new term, an important word, or words that you see onscreen. For instance, words in menus or dialog boxes appear in **bold**. Here is an example: "**Client System** sends the batches of messages to **Dispatcher Grain**, which enumerates through the batch of messages to dispatch the messages to each target grain."

> **Tips or Important Notes**
> Appear like this.

Get in touch

Feedback from our readers is always welcome.

General feedback: If you have questions about any aspect of this book, email us at customercare@packtpub.com and mention the book title in the subject of your message.

Errata: Although we have taken every care to ensure the accuracy of our content, mistakes do happen. If you have found a mistake in this book, we would be grateful if you would report this to us. Please visit www.packtpub.com/support/errata and fill in the form.

Piracy: If you come across any illegal copies of our works in any form on the internet, we would be grateful if you would provide us with the location address or website name. Please contact us at copyright@packt.com with a link to the material.

If you are interested in becoming an author: If there is a topic that you have expertise in and you are interested in either writing or contributing to a book, please visit authors. packtpub.com.

Share Your Thoughts

Once you've read *Distributed .NET with Microsoft Orleans*, we'd love to hear your thoughts! Scan the QR code below to go straight to the Amazon review page for this book and share your feedback.

https://packt.link/r/1801818975

Your review is important to us and the tech community and will help us make sure we're delivering excellent quality content.

Section 1 - Distributed Applications Architecture

Every journey begins with a first step. We will start by learning about distributed applications. These are becoming more and more common as the demand for highly scalable and available applications is increasing, and they are large and intricate. This section will provide you with a basic understanding of distributed systems by taking a quick look at the types of distributed systems and the different patterns to architect them.

In this section, we will cover the following topics:

- *Chapter 1, An Introduction to Distributed Applications*
- *Chapter 2, Cloud Architecture and Patterns for Distributed Applications*

1
An Introduction to Distributed Applications

In today's world of digital platforms, all applications are expected to be highly scalable, available, and reliable with world-class performance. Many IT companies use the term **ARP**, which stands for **availability, reliability, performance**, and have a guaranteed **service-level agreement** (**SLA**) for each. Any SLA below 99.99% (four nines) is not acceptable for an application given the challenging and highly competitive world we are in. Some mission-critical applications guarantee a higher SLA of >=99.999% (five nines). To meet increasing user demands and requirements, an application needs to be highly scalable and distributed.

Distributed applications are applications/programs that run on multiple computers and communicate with each other through a well-defined interface over a network to process data and achieve the desired output. While they appear to be one application from the end user's perspective, distributed applications typically have multiple components.

This chapter will cover the following:

- Monolithic applications versus distributed applications
- Challenges with distributed applications
- Designing your application for scalability
- Designing your application for high availability

Technical requirements

A basic understanding of Azure is all that is required to read this chapter.

Monolithic applications versus distributed applications

In the following diagram, we have a classic monolithic hotel booking application with all the UX and business processing services deployed in a single application server tightly coupled together with the database on the left side. We have a basic N-tier distributed hotel booking application with UX, business processing services, and a database all decoupled and deployed in separate servers on the right side.

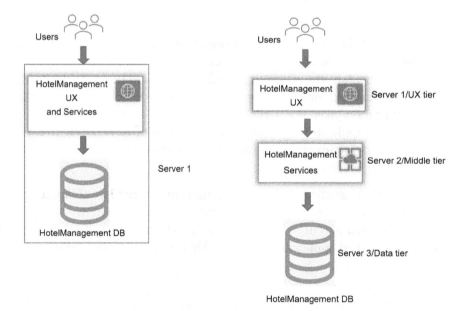

Figure 1.1 – Monolithic application (left) versus an N-tier distributed application (right)

Monolithic architecture was widely adopted 15-20 years ago, but plenty of problems arose for software engineering teams when systems grew and business needs expanded with time. Let's see some of the common issues with this approach.

Common issues with monolithic apps

Let's have a look at the scaling issues:

- In a monolithic app, there will be no option to scale up UX and services separately as they are tightly coupled. Sometimes scaling doesn't help due to conflicting needs of the resources.

- As most components use a common backend storage, there will be a possibility of locks when everyone tries to access the data at the same time leading to high latency. You can scale up but there will be physical limits to what a single instance of storage can scale.

Here are some issues associated with availability, reliability, and performance SLAs:

- Any changes in the system will need the deployment of all UX and business components, leading to downtime and low availability.

- Any non-persistent state-like session stored in a web app will be lost after every deployment. This will lead to abandoning all the workflows that were triggered by the user.

- Any bugs such as memory leaks or any security bugs in any module make all the modules vulnerable and have the potential to impact the whole system.

- Due to the highly coupled nature and the sharing of resources within modules, there will always be resource starvation or unoptimized usage of resources, leading to high latency in the system.

Lastly, let's see what the impacts are on the business and engineering team:

- The impact of a change is difficult to quantify and needs extensive testing. Hence, it slows down the rate of delivery to production. Even a small change would need the entire system to be deployed.

- Given a single highly coupled system, there will always be physical limits on collaboration across teams to deliver a feature.

- New scenarios such as mobile apps, chatbots, and analysis engines will take more effort as there are no independent reusable components and services.

- Continuous deployment is almost impossible.

Let's see how these issues are addressed in a distributed application.

N-tier distributed applications

N-tier architecture divides the application into n tiers:

- Presentation (known as the UX layer, UI layer, and the work surface)

- Business (known as the business rules layer and services layer)

- Data (known as the data storage and access layer)

Let's have a look at the advantages of a distributed application:

- These tiers can be owned, managed, and deployed separately. For example, any bug fixes or changes in the UX or service will need regression testing and deployment of only that portion.

- Multiple presentation layers, such as web, mobile, and bots, can leverage the same business and data tiers as they are decoupled.

- Better scalability: I can scale up my UX, services, and database independently. For example, in the following diagram, I have horizontally scaled out each of the tiers independently.

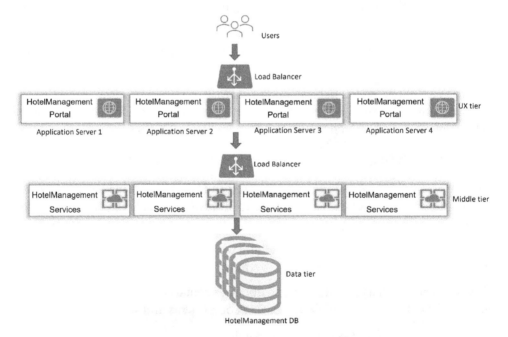

Figure 1.2 – N-tier Distributed application scaled out

- The separation of concerns has been taken care of. The presentation tier containing the user interface is separated from the services tier containing business logic, which is again separated from the data access tier containing the data store. High-level components are unaware of the low-level components consuming them. The data access tier is unaware of the services consuming it, and the services are unaware of the UX consuming them. Each service is separated based on business logic and the functionality it is supposed to provide.

- Encapsulation has been taken care of. Each component in the architecture will interact with other components through a well-defined interface and contracts. We should be able to replace any component in the diagram without worrying about its internal implementation if it adheres to the contract. The loosely coupled architecture here also helps in faster development and deployment to the market for customers. Multiple teams can work in parallel on each of their components independently. They share the contract and timelines for integration testing at the beginning of the project with each other and once internal implementation and unit tests are done, they can start with integration testing.

In this section, we discussed the advantages of distributed applications over monolithic applications and how easy it is to scale each of the tiers independently. In the next section, we will see challenges with distributed applications.

Challenges with distributed applications

When you start developing distributed applications, you will run into the following common set of challenges that all developers face:

- Design and implementation
- Data management
- Messaging

Design and implementation

The decisions made during the design and implementation phase are very important. There are several challenges, such as designing for high availability and scalability. Some design changes in the later stages of the project might incur huge costs in terms of changes to development, testing, and so on, depending on the nature of the change. Hence, it is very important to arrive at the right design choices at the beginning.

Data management

As data is spread across different regions and servers in distributed applications, there are several challenges, such as data availability, maintaining data consistency in different locations across multiple servers, optimizing your queries and data store for good performance, caching, security, and many more.

Messaging

As all components are loosely coupled in distributed applications, asynchronous messaging will be widely used for functionalities such as sending emails, uploading files, and so on. The user doesn't have to wait for these operations to be completed as these can happen asynchronously in the background and then send notifications to the user on completion. While there are several benefits, such as high performance, better scaling, and so on, there are several challenges as well with asynchronous messaging, such as handling large messages, processing messages in a defined order, idempotency, handling failed messages, and many more.

In the next section, we will see how to design applications for scalability.

Designing applications for scalability

Scalability is the ability of a system to adapt itself to handle a growing number of incoming requests successfully by increasing the resources available to the system. Scalability is measured by the total number of requests your application can process and respond to successfully. How do you know your application has reached its threshold of the maximum capacity limit? When it is busy processing current requests in the pipeline and can no longer take any incoming requests and process them successfully. Also, your application may not perform as expected, resulting in performance issues, and some requests will start to fail by timing out. At this stage, we must scale our application for business continuity. Let's look at the options available.

Vertical scaling or scaling up

Vertical scaling or scaling up means adding more resources to individual application servers and increasing the hardware capacity. Users send requests and the application processes the requests, reads/writes to the database, and sends responses back to the users. If the user base grows and the number of incoming requests becomes high, the application server will be overloaded, resulting in longer processing times and latency in responding to users. In this case, we can scale up the application server hardware to a higher hardware capacity, as shown in the following diagram.

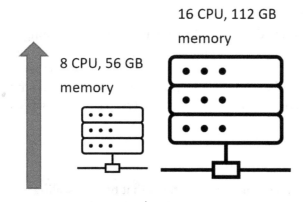

Figure 1.3 – Vertical scaling (scaling up)

Horizontal scaling or scaling out

Horizontal scaling or scaling out means adding more processing servers/machines to a system. Let's say my application is running on one server and can process up to 1,000 requests per minute. I could scale out by adding 4 more servers and could process 4,000 more requests per minute, as shown in the following screenshot.

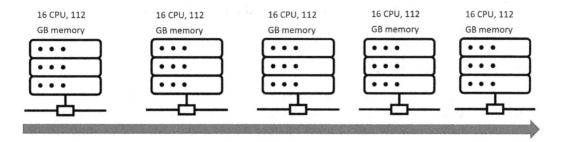

Figure 1.4 – Horizontal scaling (scaling out)

> **Tip**
>
> Having a single server is always a bottleneck beyond a certain load, no matter how many CPU cores and memory you have. That's when horizontal scaling or scaling out may help.

Load balancers

Load balancers help in increasing scalability by distributing incoming traffic to healthy servers within a region when the amount of simultaneous traffic increases. Load balancers have health probe monitors to monitor a given port on each of the servers to check the health, and if they're found to be unhealthy, the server is disabled from the load balancer and incoming traffic. When the next health probe test passes, the server is added back to the load balancer.

Caching

Caching is one of the key system design patterns that help in scaling any application, along with improving response times. Any application typically involves reading and writing data from and to a data store, which is usually a relational database such as SQL Server or a NoSQL database such as Cosmos DB. However, reading data from the database for every request is not efficient, especially when data is not changing, because databases usually persist data to disk and it's a costly operation to load the data from disk and send it back to the browser client (or device in the case of mobile/desktop applications) or user. This is where caching comes into play. Cache stores can be used as a primary source for retrieving data and falling back to the original data store only when data is not available in the cache, thus giving a faster response to the consuming application. While doing this, we also need to ensure that the cached data is expired/refreshed as and when data in the original data store is updated.

Distributed caching

As we know, in a distributed system, the data store is split across multiple servers; similarly, distributed caching is an extension of traditional caching in which cached data is stored in more than one server in a network. Before we get into distributed caching, here's a quick recap of the CAP theorem:

- **C**: Stands for consistency, meaning the data is consistent across all the nodes and has the same copy of data

- **A**: Stands for availability, meaning the system is available, and failure of one node doesn't cause the system to go down

- **P**: Stands for partition tolerant, meaning the system doesn't go down even if the communication between nodes goes down

As per the CAP theorem, any distributed system can only achieve two of the preceding principles, and as distributed systems must be partition-tolerant (P), we can only achieve either the consistency (C) of data or the high availability (A) of data.

So, distributed caching is a cache strategy in which data is stored in multiple servers/nodes/shards outside the application server. Since data is distributed across multiple servers, if one server goes down, another server can be used as a backup to retrieve data. For example, if our system wanted to cache countries, states, and cities, and if there were three caching servers in a distributed caching system, hypothetically there would be a possibility that one of the cache servers would cache countries, another one would cache states, and one would cache cities (of course, in a real-time application, data is split in a much more complex way). Also, each server would additionally act as a backup for one or more entities. So, on a high level, one type of distributed cache system looks as shown:

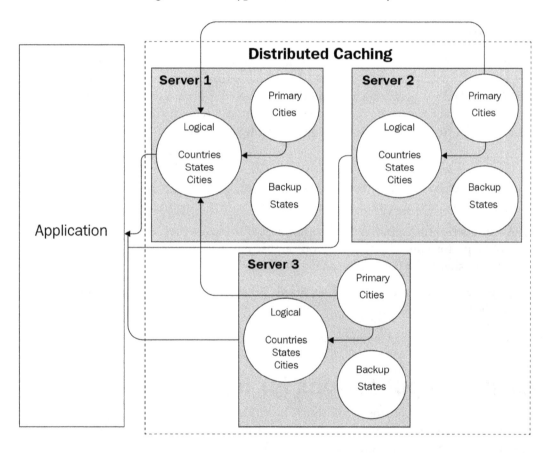

Figure 1.5 – Distributed caching high-level representation

As you can see, while reading data, it is read from the primary server, and if the primary server is not available, the caching system will fall back to the secondary server. Similarly, for writes, write operations are not complete until data is written to the primary as well as the secondary server, and until this operation is completed, read operations can be blocked, hence compromising the availability of the system. Another strategy for writes could be background synchronization, which will result in the eventual consistency of data, hence compromising the consistency of data until synchronization is completed. Going back to the CAP theorem, most distributed caching systems fall under the category of CP or AP.

Sharding

Sharding can improve scalability when storing and accessing large data from data stores. This is achieved by splitting a single data store into multiple horizontal partitions or shards. As the data is split across a cluster of databases, the system will be able to store a large amount of data and at the same time, the system can handle additional requests. We can continue to scale the system out by adding further shards.

Here are a few important considerations:

- Keep shards balanced for even load distribution. Periodically rebalance shards as data is updated and removed from each shard.

- Avoid queries that retrieve data from multiple shards as they are not efficient and cause a performance bottleneck. You can use parallel tasks to fetch data from different shards for better efficiency but it adds complexity.

- Creating a large number of smaller shards is better for load balancing than a small number of large shards.

In the next section, we will see how to design applications for high availability.

Designing applications for high availability

High availability ensures business continuity by reducing outages and disruption for customers even when some components fail in a distributed application. Let's look at some of the ways to achieve high availability.

In a scaled-out N-tier distributed application, we can add more servers to all the tiers, but I did not mention anything about Azure data centers, Azure regions, Azure Load Balancer, Azure Traffic Manager, Azure availability sets, Azure availability zones, or SQL Always On availability groups. Let's discuss each of these offerings from Microsoft Azure and see what benefits it gives to our distributed application to make it highly available.

Azure data centers

Azure data centers are physical infrastructures or buildings located all over the world where Microsoft Azure servers/VMs and services are managed.

Azure regions

An Azure region is a set of data centers connected through a dedicated low-latency network. Microsoft has 60+ Azure regions all over the world – more than any other cloud provider – from which customers can choose to deploy their applications.

Azure paired regions

An Azure paired region, as the name suggests, is a set of two regions and each region consists of a set of data centers connected through a dedicated low-latency network. The main benefit of going with paired regions is where there is a broader Azure outage affecting multiple regions, at least one region in each pair will be prioritized by Azure for quicker recovery. Planned Azure system patches and updates are rolled out sequentially to one region after another in paired regions to minimize outages or downtime in the rare case of bugs or issues with updates being rolled out.

> **Tip**
>
> You can read in detail about paired regions here: `https://docs. microsoft.com/en-us/azure/best-practices- availability-paired-regions`, which will help you understand the best practices, different regional pairs available across the globe for you to choose from, and their benefits.

Azure Traffic Manager

Azure Traffic Manager is a DNS-based load balancer to distribute traffic to internet-facing endpoints across global regions.

Traffic Manager provides a wide range of options to route traffic. Let's look at some of the frequently used routing options that can be configured in your Traffic Manager profile:

- **Priority**: This option enables you to set a primary service endpoint to which all traffic is routed and provides the option to configure backup endpoints that will take traffic when the primary endpoint is not available. This routing option is very useful in scenarios where you want to provide reliable services to your customers by having backup endpoints.

- **Weighted**: This option enables you to distribute traffic across a set of endpoints based on pre-defined weights. The weight is an integer and the higher the weight, the higher the priority. You can configure the same weight across all endpoints to distribute traffic evenly. This routing option is very useful in scenarios where you want to gradually increase the traffic to a new endpoint or provide specific weightage to certain endpoints when you are horizontally scaling up.

- **Performance**: This option enables you to distribute traffic to the "closest" endpoint for the user. The closest endpoint is not measured by geographic distance but based on the lowest network latency. Traffic Manager maintains a lookup latency table for the closest endpoint between different source IP address ranges and the Azure data center. This routing option is very useful in scenarios where you want to improve the responsiveness of your applications.

Traffic Manager provides endpoint monitoring and automatic endpoint failover as well. Let's look at important settings to be configured in your Traffic Manager profile for endpoint monitoring:

- **Protocol**: You can set HTTP, HTTPS, or TCP as the protocol that Traffic Manager can use to probe your endpoints' health. Please note that HTTPS monitoring just checks whether a certificate is present or not and doesn't check whether a certificate is valid or not.

- **Port**: You can set the port that Traffic Manager can use to send a request.

- **Expected status code ranges**: You can set success status code ranges in the format 200-299, 301-301. When these status codes are received as a response once a health check is done, Traffic Manager marks those endpoints as healthy. If you don't set anything, a default value of 200 is defined as the success status code.

- **Probing interval**: You can set an interval to specify the frequency of endpoint monitoring health check runs from Traffic Manager. You have options to set 30 seconds (normal probing) and 10 seconds (fast probing). If you don't set anything, a default value of 30 seconds is defined as the probing interval.

- **Tolerated number of failures**: You can set the total number of failures Traffic Manager can consider before making an endpoint unhealthy. You have options to set it between 0 and 9. A value of 0 means the endpoint will be marked as unhealthy for even a single failure. If you don't set anything, a default value of 3 is considered.

- **Probe timeout**: You can set the timeout value Traffic Manager can consider before making an endpoint unhealthy when no response is received. You can set the timeout value between 5 and 10 seconds when the probing interval is 30 seconds. If you don't set anything, a default value of 9 seconds is set for probe timeout.

A Traffic Manager probe initiates a GET request to the endpoint to be monitored using the protocol, port, and relative path given. If the probing agent receives a 200-OK response or any of the responses configured in the expected status code ranges, it marks the endpoint as healthy. If the response is different from any of the responses configured in the expected status code ranges or no response is received within the timeout period, the probing agent reattempts till the tolerated number of failures is reached. The endpoint is marked unhealthy once the consecutive failures count is higher than the **Tolerated number of failures** setting.

You can configure the routing and endpoint monitoring settings in your Traffic manager profile as shown in the following screenshot. The following are just sample values; you can set them based on your application's requirements.

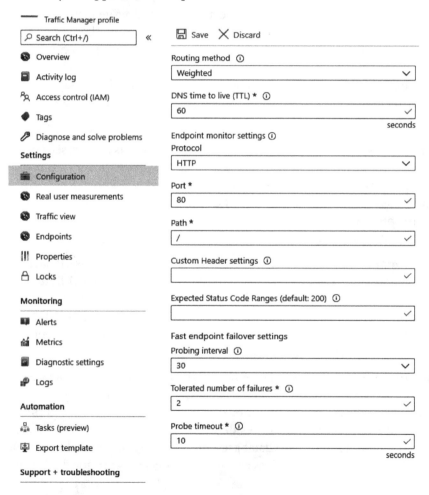

Figure 1.6 – Traffic Manager profile settings

Availability sets and availability zones

An availability set is a logical group of VMs within a data center in an Azure region and promises availability of 99.95%. They don't provide resiliency and high availability in the event of an entire data center outage.

An availability zone is made up of one or more data centers with independent power, cooling, and networking. It's a physical location within an Azure region and provides high availability (99.99%) even in the event of data center failures.

SQL Always On availability groups

SQL Always On availability groups were introduced in SQL Server 2012 to increase database availability. Availability groups support a set of read-write primary databases and one to eight sets of secondary databases to which we can fail over. These sets of databases are also called availability databases. Having primary databases and secondary databases in different Azure regions will give us high availability and resiliency against data center and Azure region failure. You can create a listener for an availability group and share that connection string for clients to connect to a database. Commit mode and failover are two important factors to consider in this list:

- **Synchronous commit mode**: In this mode, confirmations are not sent back to the client until the data is committed to a secondary database. In a way, this provides a 100% guarantee that every transaction that is committed on a given primary database has also been committed on the corresponding secondary database. Hence, this is the preferred option to sync data between databases within the same region but not for databases across regions due to latency.

- **Asynchronous commit mode**: In this mode, confirmations are sent back to the client as soon as the data is committed to the primary database without waiting to commit it to a secondary database. This mode is suitable when you want to reduce the response latency or in scenarios where primary and secondary databases are distributed over a considerable distance. Hence, this is the preferred option to sync data between databases across two different regions.

- **Automatic failover**: An automatic failover enables a secondary database to automatically transition to the primary database when the primary database becomes unavailable. Automatic failover is the preferred option when the primary database and secondary database reside within the same region with data always synchronized between the two databases. For cross region manual failover is the preferred option to avoid data loss as data is usually asynchronously committed.

Architecture for high availability

Let's extend the N-tier scaled-out distributed application to be highly available based on the options we discussed, as shown in the following screenshot.

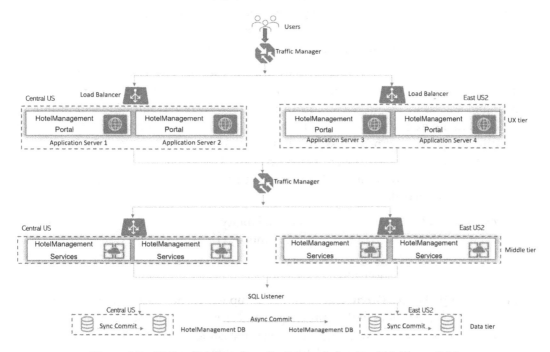

Figure 1.7 – N-tier distributed application scaled out and highly available

- **Leverage Azure paired regions** for high availability for your UX tier, middle tier, and data tier. One region will be the primary region and the other region will be the secondary region. If one region goes down, the other region will be available as a backup region. In this case, I have gone with the American regions, with the primary region as Central US and East US 2 as the secondary region. You have regional pairs available in Asia, Europe, and Africa as well. Depending upon the customer's location base, you can select regional pairs appropriately. When you combine Azure Traffic Manager with Azure Load Balancer, you get global traffic management combined with a local failover option.

- **Leverage stateless services** as any of the servers in your application can handle incoming requests and processes. Stateful services maintain contextual information during transactions and subsequent requests within a transaction need to hit the same server, hence designing for high availability and scalability becomes a challenge.

- **Leverage Active-Active mode**, which enables traffic to be routed to both regions and to load-balanced incoming requests. If one region becomes unavailable, it is automatically taken out of rotation. Active-Passive mode enables traffic to be routed to only one region at a time and would require manual failover to a secondary region when the primary region goes down, hence is not the right option for high availability unless your service is stateful and needs to maintain a sticky session where requests need to hit the same server every time for the active user session.

- **Leverage deployment of multiple instances of your service in each region**: The DNS names of these two instances are `eCommerce-CUS.cloudapp.net` and `eCommerce-EUS2.cloudapp.net`. Create a Traffic Manager profile with the name `eCommerce-trafficmanager.net` and configure it to use a weighted routing method across two endpoints, `eCommerce-CUS.cloudapp.net` and `eCommerce-EUS2.cloudapp.net`. Configure the domain name `eCommerce.com` to point to `eCommerce-trafficmanager.net` using a DNS CNAME record.

- **Leverage availability zones** to get high availability (99.99%) and resiliency against data center failures. Each of the two endpoints `eCommerce-CUS.cloudapp.net` and `eCommerce-EUS2.cloudapp.net` are configured to run on multiple servers within each region and all the servers run under the availability zone.

- **Leverage SQL Always On high availability** set up with sync commit and auto failover between databases/nodes in the same region and async commit and manual failover across the regions, as shown in the following screenshot, which is a magnified view of the database from an architecture diagram. When the application connects to a SQL availability group listener, calls will be routed to the **Node** 1 (**N1**) part of **Datacenter 1** (**DC1**), which is the primary region and primary read/write database.

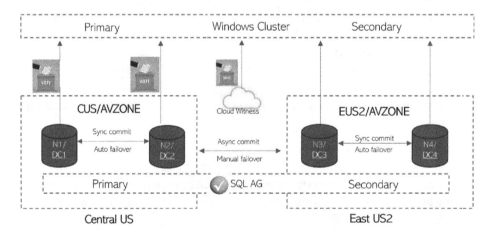

Figure 1.8 – SQL high availability setup

Let's look at two different scenarios when there is an outage and how this setup will help with high availability:

Scenarios	Mitigation	Outcome
Single-node N1 failure in Central US due to VM issue or Datacenter (DC1) outage	Auto Selfheal. Failover to Node 2 and DC2 happens automatically.	• High availability • Predictability • Less time to mitigate the issue
Multiple-node failure (persistent failure) due to VM issues or the outage of both data centers (DC1 and DC2) in Central US – a very rare scenario	Manual failover	
Single-node failure or multiple-node failure due to a VM issue or data center outages in East US2	No impact on the application as it is happening in a secondary region that the application is not pointing to.	

Summary

In this chapter, we discussed the difference between monolithic and distributed applications and why distributed applications are the way forward. We also discussed challenges with distributed applications and how to architect your distributed applications for high availability and scalability. In the next chapter, we will learn in detail about proven design patterns and principles to handle the different challenges with design, data management, and messaging in distributed applications.

Questions

1.   What are the options available to increase the scalability of the system?

   A. Vertical scaling or scaling up

   B. Horizontal scaling or scaling out

   C. Both

   D. None of the above

   Answer – C

2.   What is vertical scaling or scaling up?

   A. Vertical scaling or scaling up means adding more resources to a single application server and increasing its hardware capacity, typically achieved by increasing the capacity of the CPU or memory.

   B. Vertical scaling or scaling up means adding more processing servers/machines to a system.

   C. Vertical scaling or scaling up means adding caching to your system to avoid database calls for frequently used objects.

   D. None of the above.

   Answer – A

3.   What is an availability zone?

   A. An availability zone is a logical group of VMs within a data center in an Azure region and promises availability of 99.95%. They don't provide resiliency and high availability in the event of the outage of an entire data center.

   B. An availability zone is made up of one or more data centers with independent power, cooling, and networking. It's a physical location within an Azure region and provides high availability (99.99%) even in the case of data center failures.

   C. An availability zone is a pair of regions and each region consists of a set of data centers connected through a dedicated low-latency network.

   D. An availability zone is a DNS-based load balancer to distribute traffic to internet-facing endpoints across global regions.

   Answer – B

4. Which of the following statements is correct?

    A. SQL Always On availability groups support a set of read-write primary databases and 1 to 12 sets of secondary databases to which we can fail over.

    B. SQL Always On availability groups support a set of read-write primary databases and one to four sets of secondary databases to which we can fail over.

    C. SQL Always On availability groups support a set of read-write primary databases and one set of secondary databases to which we can fail over.

    D. SQL Always On availability groups support a set of read-write primary databases and one to eight sets of secondary databases to which we can fail over.

    Answer – D

5. Services can be deployed and scaled independently. Issues in one service will have a local impact and can be fixed by just deploying the impacted service.

    A. Domain-driven design principle

    B. Single-responsibility principle

    C. Stateless service principle

    D. Resiliency principle

    Answer – B

6. What are stateful services?

    A. Stateful services maintain contextual information during transactions and subsequent requests within a transaction need to hit the same server, hence designing for high availability and scalability becomes a challenge.

    B. Stateful services do not maintain contextual information during transactions and subsequent requests within a transaction can hit any server, hence designing for high availability and scalability is not a challenge.

    C. Stateful services are the right approach to build your services for a highly scalable distributed application.

    Answer – A

2
Cloud Architecture and Patterns for Distributed Applications

Building distributed applications in the cloud can be challenging due to the complexity involved in designing loosely coupled reusable components, asynchronous processing, maintaining data consistency and data synchronization, handling large messages in a defined order, optimizing queries and data stores for better performance, and much more. To address and overcome these challenges, there are proven design principles, cloud architecture styles, and design patterns to build world-class reliable, scalable, and secure applications. These patterns along with Orleans cross-platform framework that we will learn in upcoming chapters will help you to build robust, scalable, and distributed applications.

This chapter will cover the following:

- A primer on common design principles
- Understanding cloud architecture styles
- Understanding cloud design patterns

Technical requirements

To follow along with this chapter, you will need a basic understanding of Azure.

A primer on common design principles

All the software written in the world solves one real-world problem or another and enables us to achieve new heights. As time changes, scenarios and the expectations of specific software change. To manage that change and deal with various aspects of software, engineers developed and evolved multiple programming paradigms, frameworks, tools, techniques, processes, and principles. A common set of principles and patterns being practiced, that were proven useful over time, becomes the guiding star for engineers to collaborate and build quality software. In this section, we will discuss some of the common design principles.

Design principles are a set of high-level abstract guidelines to be followed while designing applications for any programming language. They do not provide implementation guidelines.

Design principles

Techniques become principles if they are widely accepted, practiced, and proven useful in any industry. Those principles become solutions to make software designs more understandable, flexible, and maintainable. We will primarily cover commonly used SOLID, KISS, and DRY design principles in this section.

SOLID

SOLID principles are a subset of many principles promoted by American software engineer and instructor Robert C. Martin. These principles have become de facto standard principles in the **object-oriented programming** (**OOP**) world and have formed a core philosophy for other methodologies and paradigms.

SOLID is the acronym for the following five principles:

1. **Single responsibility principle (SRP)**: An entity such as a software module should only have a single responsibility. You should avoid giving one entity multiple responsibilities.

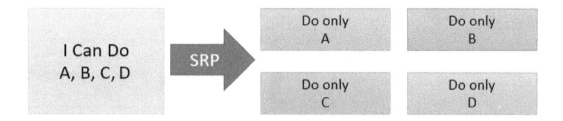

Figure 2.1 – Single responsibility principle (SRP)

2. **Open closed principle (OCP)**: Software entities such as classes, modules, and functions should be designed in a way that they should be open for extension but closed for modification. We should be able to extend the behavior of entities without modifying their source code. This avoids regression testing on existing behaviors as there will always be extensions of existing entities and the extension is the only thing that needs to be tested.

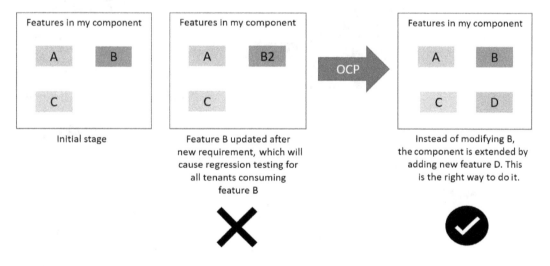

Figure 2.2 – Open closed principle (OCP)

3. **Liskov substitution principle (LSP)**: Parent or base class instances should be replaceable with instances of their derived class or subtypes without altering the sanity of the program.

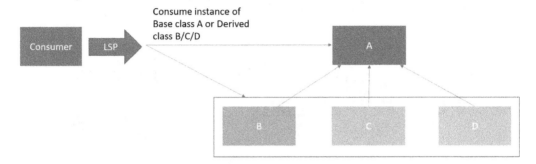

Figure 2.3 – Liskov substitution principle (LSP)

4. **Interface segregation principle (ISP)**: Instead of one common, large interface, plan for multiple scenario-specific interfaces for better decoupling and change management.

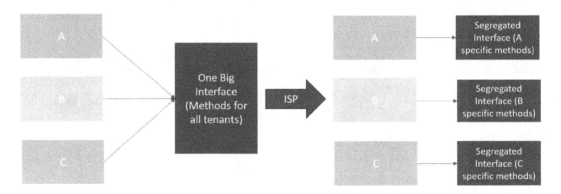

Figure 2.4 – Interface segregation principle (ISP)

5. **Dependency inversion principle (DIP)**: You should avoid taking a direct dependency on a concrete implementation. Instead, you should depend on abstractions as much as possible. Higher-level modules shouldn't depend on the lower-level modules and module dependencies should be based on abstract interfaces rather than the direct reference of concrete implementations.

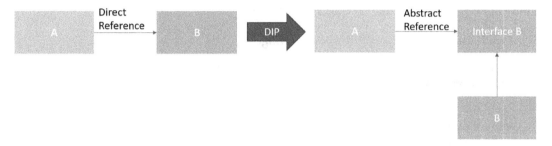

Figure 2.5 – Dependency inversion principle (DIP)

It is very common to observe that these principles are closely related to each other and empower each other in one way or another. You should not limit these principles only to objects; they can be leveraged in architecture to design components and relationships between them. Let's look at the **Don't Repeat Yourself (DRY)** design principle in the next section.

Don't Repeat Yourself (DRY)

The system should be designed in such a way that the implementation of a feature or a pattern should not be repeated in multiple places. This would result in a maintenance overhead as changes in requirements would result in modifications in multiple places. If you failed to update in one place by mistake, the behavior of the system would become inconsistent. Rather, the feature should be wrapped in a package and should be reused in all places. In the case of a database, you should look at data normalization to reduce redundancy.

This strategy helps in reducing redundancy and promoting reuse. This principle helps with the organizational culture as well, encouraging more collaboration by helping others and leveraging help from others.

Purposeful repetition can be fine. In cases of defining contracts for similar pieces of data across multiple boundaries, you would not necessarily want to rely on some shared contract across all boundaries – as boundaries' needs may change over time. Additionally, there may be cases where data is needed in one context, but not another, so having multiple records that represent *almost* the same data may be desirable over "one object to rule them all" where some properties go unused, depending on the context.

Keep it simple, stupid (KISS)

A system should be designed simply by avoiding complicated designs, algorithms, new untried technologies, and so on. Instead, focus on leveraging the right OOP concepts and reusing proven patterns and principles. Use complex designs and code only if they are necessary and add value to the implementation.

When you keep it simple, you will do the following:

- Avoid mistakes while designing/developing.

- Keep the train running (there is always a team whose job is to maintain the system and they are not always the ones who developed the system in the first place).

- Read and understand the code easily (from the perspective of someone new to the team or someone trying to reuse the complex designs and code, or you could be looking at your system design and code after many months or years).

- Change management will be less error-prone.

With this, we are done with the primer on common design principles, where we learned about SOLID, DRY, and KISS. In the next section, let's look at cloud architecture styles that we will be leveraging to build a distributed application using Orleans.

Understanding cloud architecture styles

Microservices architecture and event driven architecture are commonly used cloud architecture styles while designing distributed applications. First and foremost, the goal of any architecture is to support business needs with minimal effort (time and resources). Businesses want software to enable them, instead of acting as a bottleneck. Let's look at microservices architecture and event-driven architecture to understand their benefits.

Microservices architecture

Microservices architecture consists of small, loosely coupled, independent, and autonomous services. Each service in the following diagram implements a business capability adhering to the single responsibility pattern.

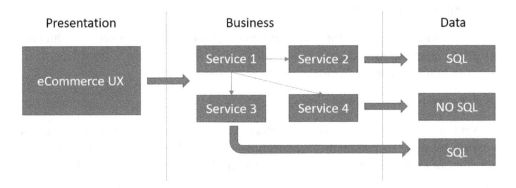

Figure 2.6 – Microservices architecture

Let's see the benefits:

- Services can be deployed and scaled independently.

- Services provide fault isolation as an issue in one service will have a local impact and can be fixed by just deploying the impacted service.

- Services are independent and there is no compulsion to share technology or frameworks.

- Services communicate with each other via well-defined APIs or messaging systems.

- Services can be owned by small independent teams and can have their own cycle.

- Services are responsible for managing their own data stores. Scenarios demanding lower latency can be optimized by bringing in a cache or high-performing NoSQL stores.

Some factors to be considered are the following:

- As the number of services increases, communications and integration testing between services can become complex.

- Debugging is challenging as each service will have its own telemetry and needs to have end-to-end tracing enabled.

Event-driven architecture

Event-driven architecture allows for near real-time communication between producers and consumers, enabling the data to be consumed in the form of events. Producers and consumers are totally decoupled from each other.

The principle specifications are as follows:

- In event-driven architecture, communication, which is generally known as Pub-Sub between modules, is primarily asynchronous and is achieved via events. The structure of events is the only contract between them, which will be exchanged.

- There can be multiple consumers of the same event taking care of specific operations, and ideally not even aware of each other. Producers can continuously push events without worrying about the availability of the consumers.

- The publisher publishes events via a messaging infrastructure such as queues and Service Buses. Once an event is published, the messaging infrastructure is responsible for sending the event to eligible subscribers.

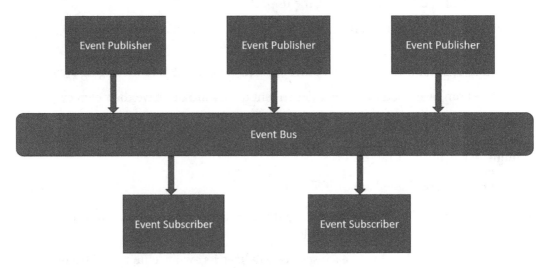

Figure 2.7 – Event-driven architecture

This is best suited for scenarios that are asynchronous in nature. For example, long-running operations can be enqueued for processing. A client might poll for the status or even act as a subscriber for an event.

With this, we are done with common cloud architectures for building distributed applications. Next, we will look at some of the common design patterns that will help in building a distributed application using Orleans in the cloud.

Understanding cloud design patterns

Patterns are low-level specific implementation guidelines that are proven reusable solutions for recurring problems. Each of the following patterns can be implemented in Microsoft Azure to build reliable, scalable, and secure distributed applications.

The gateway aggregation pattern

The gateway aggregation pattern aggregates multiple client requests targeting multiple backend services in a single client request, dispatches the requests to various backend services, then aggregates the responses received from all backend services into one response and sends it back to the client.

Usage

This pattern is very useful to reduce the chattiness between the client and multiple backend services, which can impact the application's performance. This is a very common problem in microservices architecture as well as mobile clients, where this pattern can be leveraged to reduce the number of calls to backend services. As shown in the following screenshot, the client sends an aggregated request to the gateway (1). The gateway sends a request to each service and receives a response (2, 3, 4, 5, 6, 7). The gateway aggregates the responses and sends a response back to the client (8).

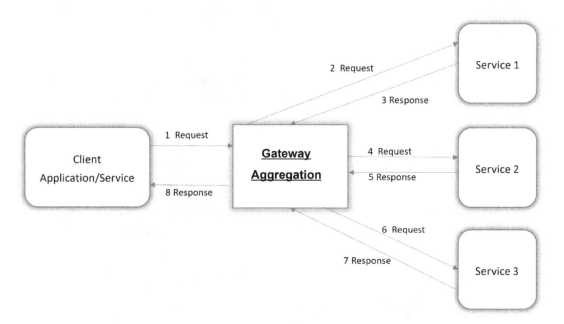

Figure 2.8 – Gateway aggregation pattern

Factors to be considered

- Ensure the gateway is highly available, scalable, resilient, and has timeouts, with retries designed properly to avoid a single point of failure.

- Ensure the gateway is not causing an increase in the response time for clients. Look at implementing parallel tasks to call each backend service instead of sequential requests.

The CQRS pattern

The **Command and Query Responsibility Segregation (CQRS)** pattern separates write operations from read operations for data stores. The responsibility is clearly segregated between the write model – a.k.a. the *command* model – and the read model – a.k.a. the *query* model – instead of the traditional one data model for all operations.

Usage

This pattern is very useful in applications where you want to separate the loads on read and write and scale each of them independently depending upon the expected traffic, especially when the load varies too much between read and write operations. In the following diagram, the traditional way of accessing the data model is on the left and the CQRS pattern is on the right. The **Ordering Service** that supports all read and write operations from the same data model is split into two services – the **Order Fetch Service** for read operations using the query model, and the **Order Transaction Service** for write operations using the command model. We can scale up the **Order Fetch Service** and **Order Transaction Service** independently depending upon the anticipated load. The query model will only have a query to read from the data store and the command model will have commands to update the data store. This pattern also helps in achieving the single responsibility design principle.

Figure 2.9 – Non-CQRS (left) versus CQRS (right)

You can also split the data store into a read data store and a write data store as shown in the following diagram. You need to make sure both data stores are in sync. If you can recall the SQL high availability setup that we discussed in *Chapter 1, An Introduction to Distributed Applications*, the primary SQL node can be used as a write store and the secondary SQL node can be used as a read store.

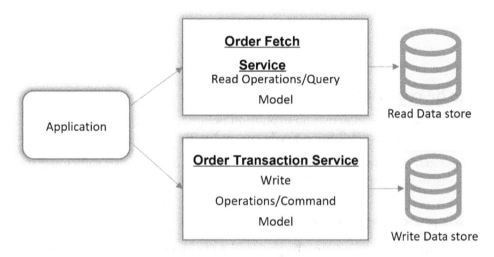

Figure 2.10 – CQRS pattern with a separate read and write store

Factors to be considered

Changes made in the write data store should be synced to the read data store to avoid stale read data and to ensure data consistency.

The cache-aside pattern

The cache-aside pattern ensures cached data is up to date and consistent with data in the data store for applications to leverage the cache in a fast way and avoid latency due to data store calls every time.

Usage

This pattern is very useful in applications where you want to improve the performance by making repeatedly accessed and expensive to read data available in the cache. As shown in the following diagram, we first must check if the data requested is available in the cache. If it's available, we can return it back, otherwise, we need to make a call to the data store to fetch and store it in the cache and then return it.

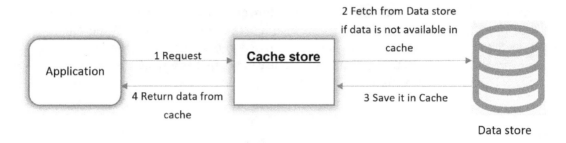

Figure 2.11 – Cache-aside pattern

Factors to be considered

- For this pattern to be effective, an expiration policy and data eviction need to be considered based on application needs to avoid retrieving data from the data store.

- Changes made in the data store can be synced to the cache additionally to ensure data consistency and to avoid data not being present in the data store.

The priority queue pattern

The priority queue pattern prioritizes the processing of requests with a higher priority than those with a lower priority.

Usage

This pattern is very useful in scenarios where you have varying SLAs based on clients or events to process the messages and complete the message. For example, an order confirmed and invoiced notification needs to be sent immediately, whereas an order feedback message can wait for some time. As shown in the following diagram, the application posting a message can assign a priority for each message based on the event type or client type and send it. The events notification service can internally have a priority queue to process the messages, with higher-priority messages first, and send them to subscribers.

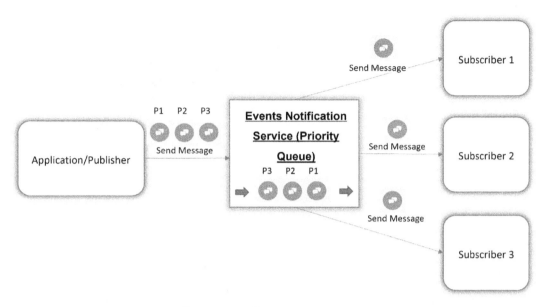

Figure 2.12 – Priority queue pattern

Factors to be considered

Ensure you have a mechanism to cancel or suspend a task that's handling low-priority message processing when a higher-priority message becomes available.

The external configuration store pattern

The external configuration store pattern moves all the configuration information required for your application to a centralized location for better control and management of the configuration data. This also makes it possible to share the configuration data across multiple applications instead of limiting the configuration to a single application.

Usage

This pattern is very useful to share common configuration settings such as database connection strings, URLs, or endpoints of external services across multiple applications. As shown in the following diagram, multiple applications can access this common configuration data from an external configuration store.

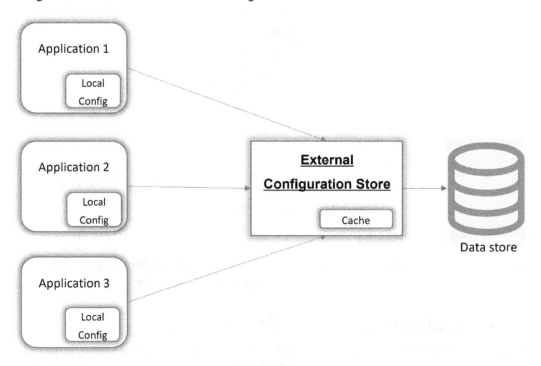

Figure 2.13 – External configuration store pattern

This pattern is also very useful when you don't want to redeploy or restart your application after configuration updates resulting in downtime, which is the case when you have a configuration local to the application and when the configuration is packaged and deployed along with the application.

Factors to be considered

- Ensure that the configuration store interface can expose data in different formats such as key-value pairs and collections.

- Ensure that the configuration store is highly available, secure, and offers good performance without latency.

- Ensure testing is performed for all impacted tenants whenever the centrally stored configuration is updated.

The pipes and filters pattern

The pipes and filters pattern has multiple components called as filters connected to each other, where each filter performs the independent task of processing the input and sending the output to the next filter in the processing pipeline via pipes.

Usage

This pattern is very useful when the application must perform several tasks to process the request and send the response to the target. Instead of one monolithic module performing all tasks, breaking them down into filters as shown in the following diagram will help in reusing the same filter elsewhere in the application if similar processing is required. A distributed application can break down its processing into multiple separate tasks, where each task is performed by a filter.

Figure 2.14 – Pipes and filters pattern

Factors to be considered

- Ensure the data flowing in the pipeline between filters is not lost by using a resilient infrastructure.

- Ensure that the filters in the pipeline can detect duplicate messages.

Summary

In this chapter, we discussed SOLID design principles, microservices and event-driven cloud architecture styles, and different messaging and data management patterns such as gateway aggregation, CQRS, cache-aside, priority queues, and external configuration stores. These are important, frequently used principles and patterns for building distributed applications.

In the next chapter, we will learn how Microsoft Orleans as a tool helps in building scalable distributed applications. We will also learn about the key architectural decisions and trade-offs made by Orleans, which make it easy for developers to build cloud-native distributed applications.

Questions

1. What is the Liskov substitution principle?

    A. Base class instances should be replaceable with instances of their derived type.

    B. Derived class instances should be replaceable with instances of their base type.

    C. Designing for generics that can work with any data type.

    Answer – A

2. What is the single responsibility principle?

    A. Instead of one common large interface, plan for multiple scenario-specific interfaces for better decoupling and change management.

    B. You should avoid taking a direct dependency on a concrete implementation; instead, you should depend on abstractions as much as possible.

    C. An entity should only have a single responsibility. You should avoid giving one entity multiple responsibilities.

    D. Entities should be designed in such a way that they should be open for extension but closed for modification.

    Answer – C

3. What is the open closed principle?

    A. Open to modification, but closed for extension

    B. Open to extension, but closed for modification

    C. Open to composition, but closed for extension

    D. Open to abstraction, but closed for inheritance

    Answer – B

4. What are the characteristics of microservices?

    A. Services can be deployed and scaled independently.

    B. Services communicate with each other via well-defined APIs or a messaging system.

    C. Services are independent and there's no compulsion to share technology or a framework.

    D. All the above.

    Answer – D

5. Which of the following factors need to be considered for the gateway aggregation pattern?

    A. Ensure the gateway is not causing an increase in the response time for clients.

    B. Ensure the gateway is highly available, scalable, resilient, and has timeouts, and retries are designed properly to avoid a single point of failure.

    C. Ensure the gateway has parallel tasks to call each backend service instead of sequential requests.

    D. All the above.

    Answer – D

6. Which of the following patterns help in data management?

    A. CQRS

    B. Ambassador

    C. Gateway

    D. All the above

    Answer – A

7. Which of the following patterns helps in asynchronous messaging between a decoupled sender and consumer?

    A. CQRS

    B. Cache-Aside

    C. Publisher-Subscriber

    D. None of the above

    Answer – C

8. How is the external configuration store pattern useful?

    A. This pattern is very useful to reduce the chattiness between the client and multiple backend services, which can impact the application's performance.

    B. This pattern is very useful to share common configuration settings such as database connection strings, URLs, or endpoints of external services across multiple applications.

    C. This pattern is very useful in legacy applications where enhancement is very difficult, be it common tasks such as authentication, authorization, logging, or monitoring.

    D. All the above.

    Answer – B

Section 2 - Working with Microsoft Orleans

This section discusses the motivation behind the Microsoft Orleans framework and its most important design choices. In this section, we'll learn about Orleans concepts such as grains, silos, state persistence, timers, and reminders. Throughout the process, we'll apply everything we've learned to create a fictitious hotel management application called Distel.

In this section, we will cover the following topics:

- *Chapter 3, Introduction to Microsoft Orleans*
- *Chapter 4, Understanding Grains and Silos*
- *Chapter 5, Persistence in Grains*
- *Chapter 6, Scheduling and Notifying in Orleans*
- *Chapter 7, Engineering Fundamentals in Orleans*

3
Introduction to Microsoft Orleans

By now, you know what distributed applications are and the challenges we face in building them to make them highly available and massively scalable distributed applications with the least latency possible. Distributed systems have the potential for more reliability because of their distributed nature. Well architected distributed applications provide the ability to scale the system quickly in response to the application's demands. This advantage of distributed systems comes with a cost: they are significantly complex to architect, build, and debug. Technology and engineering practices play a key role in building reliable systems. They demand highly skilled engineers. This presents a need for tooling that takes care of the plumbing required and helps the developer to focus on business logic. In the previous chapters, we learned about building distributed applications and different patterns of building distributed applications. In this chapter, we will learn how Microsoft Orleans as a tool helps in building scalable distributed applications. We will also learn about the key architectural decisions and trade-offs of Orleans to make it easy for the developer to build cloud-native distributed applications.

In this chapter, we will learn about the basic building blocks of Orleans:

- The actor model for distributed systems
- The origin of Orleans
- The Orleans runtime
- Orleans key design choices
- Why should I learn about Orleans?
- When to choose Orleans

Understanding the actor model for distributed systems

In this section, we will learn about the motivation behind the building of Microsoft Orleans and the different trade-offs that Microsoft Research took in defining its features.

In a traditional three-tier system, shown in the following figure, we will have a stateless presentation tier and a stateless middle tier or service layer (business tier). The storage layer (or the third tier) presents a bottleneck for scalability. In this system, the services can be scaled horizontally, so there is a possibility that several clients will attempt to access shared resources concurrently, for example, a record in a database table. In *Chapter 1, An Introduction to Distributed Applications*, we saw that we can scale the database horizontally too. With the horizontal scaling of databases, we need to deal with the challenges of maintaining data consistency across data centers. Pessimistic locking is an approach we can take to ensure data consistency. We can also use optimistic concurrency in conjunction with database transactions, that is, performing a read and update operation within a single ACID transaction. Correctness, performance, and robustness are difficult to achieve in the face of concurrency. Distribution, high throughput, and low latency make it even harder. In Microsoft service architecture, every service is independent. There will be cases of one microservice calling another microservice to serve a request, for example, a hotel service calling a reservations service for availability details. These service calls are termed horizontal calls. We need to write extra code as per the communication channel to handle transactions, latency, and failures in such scenarios. In the following architectural diagram, the presentation tier on the browser/laptop/mobile client accesses the business tier via the internet. The business tier serves these requests from the presentation tier by fetching the relevant data from the storage tier.

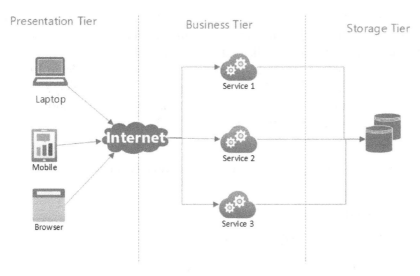

Figure 3.1 – Three-tier architecture

With the introduction of the cache layer before the storage tier, as shown in the following figure, instead of going to the datastore every time, we will be able to serve requests faster by fetching them from the cache for subsequent requests. This will address the latency problem. But the system has another problem. With the introduction of the cache, the data semantics and concurrency control are lost. This means that the data stored in the cache may be flattened and stored as key-value pairs, but it will not have the same semantics as the data stored in the datastore. With this, we will lose the data semantics. It will be even more challenging to maintain the data consistency and application semantics in situations where a single API call touches multiple entities. Here too, we need to write extra code for horizontal calls.

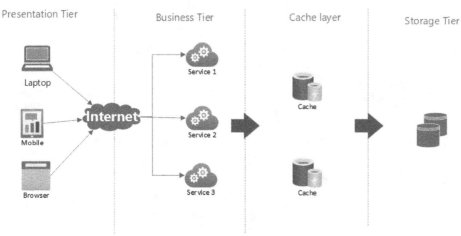

Figure 3.2 – Three-tier architecture with the cache layer

As we saw in this chapter and in the previous chapters, common programming practices are not satisfactory for the needs of modern distributed systems. There are many challenges thrown at the developer while scaling applications. To address these problems, Carl Hewitt, Peter Bishop, and Richard Steiger proposed the actor model in 1973. Their invention was to create message-passing systems. The same idea of message passing was brought into languages such as Smalltalk and Erlang and programming models such as Akka, Orleans, and Reliable Actors from Azure Service Fabric. Web workers in JavaScript also use the same core concept of message passing.

Actors are primitives of concurrent computing. The fundamental concepts of actors are as follows:

- Encapsulate the state.

- Receive and send messages from/to other actors.

- Create other actors.

Using the actor model as a stateful business layer as shown in the following figure, we can solve the concurrency control as the state is maintained in the actors and they are single-threaded. We will still be able to achieve high performance for the cache as the state is persisted in the actors without losing the data semantics. An actor can communicate with other actors and it does not have to worry about the communication channel. A `User` actor, for example, can interact with a `Cart` actor without understanding the details of the communication mechanism.

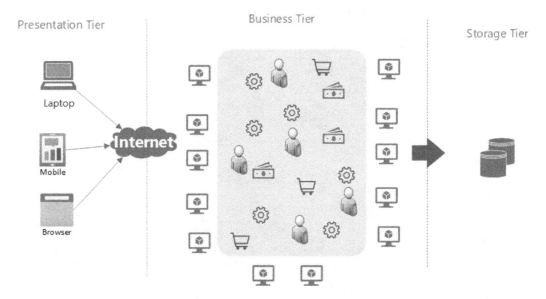

Figure 3.3 – Actor model as stateful business layer

> **Tip**
>
> You can learn more about the actor model and the principles behind it here:
> `https://en.wikipedia.org/wiki/Actor_model`.

We have seen the challenges faced by developers to build distributed applications and how the actor model addresses them. In the next section, we will learn about the origin of Microsoft Orleans and its design principles to ease the developer's effort in building distributed applications.

The origin of Microsoft Orleans

Though the actor model helps in building distributed applications, there is a steep learning curve. The eXtreme Computing Group of Microsoft Research published a white paper (`https://www.microsoft.com/en-us/research/project/orleans-virtual-actors/`) on the actor model in 2010. The primary focus of this white paper is to simplify distributed computing and allow non-experts to write efficient, scalable, and reliable distributed systems. Microsoft Orleans is a runtime and framework built based on the white paper. In Orleans, the actor is termed a **grain**. The life cycle of a grain is managed by Orleans. Orleans actors are virtually always present. We just get the reference to the grain activation to interact with. In the case of other actor implementations, the developer has to handle the creation of the actor, placement, and recovery from failures. Orleans makes it simple to build distributed applications by abstracting all the complexity. The developer can build distributed applications with the well-known object-oriented programming model.

Microsoft made Orleans an open source project in 2015. Since then, the open source developer community has embraced it and has both utilized Orleans and helped improve it by making contributions. Microsoft Orleans is time-tested. It has been running in production for over 8 years now. Orleans is used in different Microsoft products with different use cases. Azure PlayFab, Xbox Game Studios, Azure Active Directory, Azure IoT Digital Twins, Azure Quantum, Skype, Azure Machine Learning, and Dynamics 365 Fraud Detection are some examples of Microsoft products leveraging Orleans. The most well-known use of Orleans is within the Halo game series, with over 11 million players, 1.5 billion games played, and 270 million hours of game play.

So far, we have learned about the motivation behind the development of Microsoft Orleans. In the next section, we will detail the Orleans runtime.

The Orleans runtime

We learned in the previous section that the grains in Orleans are virtually always present. They cannot be created or destroyed. Here, the heavy lifting is done by the Orleans runtime. The Orleans runtime ensures that a grain is always available to serve commands. When a request comes to a grain that is currently not instantiated, the Orleans runtime will automatically activate the grain on a server and initialize it to serve the request. The Orleans runtime will not instantiate the grain if there are no requests pending on it. If the server where the grain is currently instantiated fails, the runtime will automatically create the grain on another available server on the next invocation. Here, the developer need not keep track of the re-creation of a failing actor, unlike other actor programming models. The Orleans runtime will take care of the lifetime management for the grain. While activating or deactivating a grain, the Orleans runtime will also notify the application, so that we will have the flexibility to perform any task when an activation or deactivation event occurs. We will learn more about this in upcoming chapters.

Orleans runs on a cluster of servers or nodes in a data center. In Orleans, a node is referred to as a silo. The Orleans runtime that will be running on these nodes activates and hosts grains. The three key subsystems that constitute the runtime are messaging, hosting, and execution.

The messaging system

As the name suggests, the messaging system handles the communication between nodes in the cluster and grains. Every node in the cluster communicates with other nodes using a single dedicated TCP connection per node. The messaging system also manages the communication threads to multiplex the messages between grains hosted on silos. When a message is sent to a grain, the messaging system first consults the hosting system to get the grain activation and then the parameters are serialized and passed on to the activation through an already opened socket to that server.

The hosting system

The hosting system manages the placement of a grain upon activation. It also manages the life cycle of grains. It is responsible for the activation or deactivation of grains. It deactivates the grains that are not in use for a specified time. The Orleans runtime also provides a way to customize the grain location, which we will learn about in upcoming chapters. The hosting system keeps track of all the grain activations by maintaining a distributed directory. The hosting system also takes care of the local resources in the server. It is responsible for deactivating a grain if it is idle for a configured time.

The execution system

The execution system runs the grain's application code on a single thread. When one grain calls another grain, the executing system converts the method call to a message and passes that onto the messaging system. The requests received by a grain are queued and processed sequentially in a single thread.

In this section, we have learned about the major subsystems that constitute the Orleans runtime. In the next section, we will learn about the key architectural choices made by the Orleans team to make distributed application development easy.

Orleans design choices

With the main guiding principle being to enable the simple programming model without sacrificing performance, the Orleans team made some design choices. In this section, we will learn about the key design choices made on the placement of the grain, state isolation, how asynchrony is achieved, and a few more.

Serialization

Serializing and deserializing complex objects is part of any inter-process communication. Mostly, this process is hidden from developers, but its efficiency has a direct impact on the overall performance of the system. The Orleans runtime serializes the data passed in grain requests and response messages. It also serializes persistent state objects. The serialization and deserialization of objects in Orleans are done by an built-in highly optimized serialization subsystem. This is achieved by generating the serialization code at the time of compilation. The two important features of the Orleans serialization framework are as follows:

- *Dynamic types and arbitrary polymorphism*: These maintain the dynamic nature of the actual data types. For example, if a grain interface is declared to accept `IList` but at runtime, the caller passes `SortedList`, the receiver will indeed get `SortedList`.

- *Maintained object identities*: If the same object is passed in more than once in a message, Orleans will serialize it only once. On the receiver side, all the references will be restored without any deviations. All the references of the object will point to the same object after deserialization. For example, if a graph is passed as an argument by a sender, all the nodes in the graph are serialized only once. On the receiver side post deserialization, the graph will look exactly the same as the sender. This is possible by preserving the object's identity.

To generate serializers at the time of compilation, Orleans code generators scan all the types in those assemblies where the Orleans core library is directly referenced and then generates the serialization code for the types directly referenced in grain interfaces, grain state, or any types marked with the `[Serializable]` attribute. There are custom serializers such as `Orleans.Serialization.ProtobufSerializer`, `Orleans.Serialization.BondSerializer`, and `Orleans.Serialization.OrleansJsonSerializer` in addition to the Orleans default serializer, which developers can use if needed. There is also a provision to implement a custom serializer by implementing `IExternalSerializer`.

Strong isolation

Orleans grains do not share state with others and they are isolated from each other. Grains can only communicate with others by sending messages. This is as per the actor model principle of communication. The arguments passed in the message call are deep copied by the serializer synchronously. Even if the sender and receiver are on the same node, messages are sent by serialization to guarantee immutability. To reduce the cost of deep copying, Orleans uses highly optimized serialization as described in the previous section. In addition to this, to optimize the object serialization even further, the application can specify that it will not mutate an argument by wrapping the type using `Immutable<T>` or adding the `[Immutable]` attribute to the argument in the method signature. This tells the runtime that it is safe to pass the argument without deep copying.

Asynchrony

Grains are single-threaded; calls may need to wait for other calls to complete before they can execute. So all grain operations are asynchronous. Orleans enforces the asynchronous programming model of .NET. All the methods in the grains must return `Task` or `Task<T>` or `ValueTask` or `ValueTask<T>`. The use of the asynchronous model prevents the application code from holding a thread while waiting for a result and ensures that throughput is minimally impacted.

The distributed directory

Orleans maintains a distributed directory to track the placement of grains, unlike other distributed systems that use a static placement scheme, such as consistent hashing. This gives the Orleans runtime flexibility in managing system resources by placing and moving grains as the load on the system changes. The Orleans distributed directory is implemented as a one-hop distributed hash table where each node in the cluster holds a partition of the directory. Each record in the partition maps a grain to a process in the cluster. In the case of a new activation, a registration request is sent to an appropriate partition, and similarly in the case of deactivation, to free up resources. Orleans uses a consistent hashing algorithm to map grain IDs to directory partitions.

Maintaining the distributed directory comes with an overhead of finding the placement of the grain with every message. To address this, the Orleans runtime maintains the local cache of recently resolved actor activation mappings on all the nodes.

Eventual consistency

In a happy path scenario, Orleans guarantees that there will be only one grain active at a given time. However, in case of failures, having a grain activated only once happens eventually. The membership will be in flux when one of the nodes in the cluster failed but its failure has not been communicated to all the other nodes. During this period, there is a possibility of two activations of the same grain. However, eventually, when the membership gets settled, one of the activations will be dropped from the directory and a message will be sent to the respective node to deactivate it. This is done to ensure that applications can run even when membership is in an ambiguous state. This trade-off is made in Orleans in favor of availability over consistency. In those scenarios where eventual consistency won't work, the application can rely on external persistent storage to provide stronger data consistency. We will look at this in upcoming chapters.

Message guarantees

By default, Orleans messaging is configured to deliver messages at most once. Every message delivery in Orleans has a configured timeout. If there is no response within the configured time, the caller will receive a timeout exception. This allows developers to implement at-least-once messaging. If the developer decides to implement retries, either by configuring Orleans to retry upon timeout or by implementing application logic to retry, the message will be delivered eventually. This will result in at-least-once delivery, that is, the message might be delivered more than once.

Reliability

One of the key measures of any distributed application is reliability. You may recall the learnings from *Chapter 1, An Introduction to Distributed Applications*. The Orleans runtime takes care of all the reliability aspects required for a distributed application and eases the developer's effort. The Orleans runtime self-heals itself in the event of server failures using the membership mechanism. Servers detect failures automatically via frequent heartbeats and agree on the membership view eventually. The time it takes for the system to reach a state of convergence is determined by the failure detection parameters. When a node learns about the failure node, it scans the local directory partition and grain activation catalog and clears the activations pertaining to failure nodes. This won't impact the availability of those actors from the failing node; they will be activated on a healthy node when they receive the next request. This may or may not impact the state. Orleans does not implement the checkpointing strategy to persist the state of the grain. It is up to the application developer to decide on the best mechanism to persist the state. In some scenarios, developers might persist the state for every change in the in-memory state. We will never lose the state here, but it will have an impact on performance. Another approach is to persist the state periodically.

Cooperative multitasking

Preemptive and cooperative multitasking are the two types of multitasking. Cooperative multitasking means that once a task has begun execution, it continues uninterrupted until it is finished or yields execution. In preemptive multitasking, the scheduler initiates a context switch from one executing process to another at regular intervals or when the task signals that execution can be yielded, as shown in the following diagram.

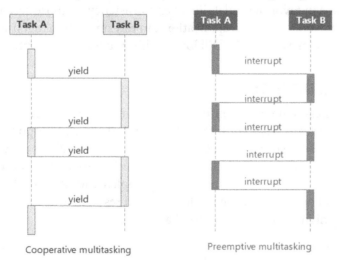

Figure 3.4 – Preemptive versus cooperative multitasking

The Orleans runtime uses cooperative multi-tasking to schedule the execution turn. Orleans schedules execution on the .NET thread pool. As a consequence of Orleans using the .NET Task Parallel Library, it supports cooperative multitasking. The downside of using cooperative multitasking is that a poorly performing grain can degrade the entire system's performance. Since .NET uses thread pools and they are preemptible, the performance degradation due to long-running tasks is not quite severe. Orleans also provides monitoring and notifications to troubleshoot such scenarios.

Single threading

Within each grain activation, Orleans ensures that only one thread is active at any given moment. As a result, several threads can never access the activation state at the same time, eliminating race situations and eliminating the need for locks and other synchronization primitives. The execution subsystem provides this assurance without the need for per-activation threads. While single-threading limits an individual activation's performance, the parallelism across numerous grain activations addressing diverse requests is more than enough to make efficient use of the available CPU resources and improves overall system responsiveness and throughput.

So far, we have learned about the important architectural choices behind the Microsoft Orleans platform. In the next section, we will learn about why we should consider Orleans while building a distributed application.

Why should I learn about Orleans?

When we plan to learn about or use a new technology, it is quite natural to compare it with other similar technologies. Orleans is often compared with Akka.NET and Erlang actors. Microsoft Orleans, with its design choices, is an actor implementation, termed a virtual actor. The trade-off design choices made, as we learned in the previous section, will ease the developer's effort in building distributed applications. In addition to this, the most important thing to note is Orleans embraces the well-known object-oriented programming model. Orleans is often termed distributed .NET as it extends familiar .NET concepts such as objects, interfaces, async/await, and try/catch and extends them to distributed context. For object-oriented programmers, the learning curve is smooth with Orleans when compared with other distributed application development frameworks.

Orleans provides a programming paradigm that smoothly blends non-distributed techniques and developer abilities, allowing scaling out beyond the constraints of a single machine without having to deal with the challenges of developing a distributed application. This is accomplished by deciding on a set of implementation options and then providing a limited set of tools to the user that can be used without having to understand the underlying technology.

The objective of Orleans is to hide the actor-inspired implementation details and provide a painless "elastic object" programming approach that does not require distributed programming skills.

According to Carl Hewitt's actor definition, "Actors create a finite number of new actors." Orleans grains always exist virtually. According to Reuben Bond (chief architect and open source contributor to Orleans), Orleans grains are cloud-native objects as they represent the entities of distributed applications. Some developers also term them digital twins as they represent a digital counterpart (state and actions) or some real-world entities.

Another important aspect to consider is that Orleans is platform-agnostic. The application can run on any platform that runs .NET applications. It can run on Kubernetes, the Azure cloud platform, AWS, GCP, and on any other classic cloud platforms.

It is time-tested. As was said earlier in this chapter, there are many products within and outside Microsoft that are built using Orleans.

Wow! This is exciting. Can I use Orleans to build any application? The answer is no. Let's see why in the next section.

When to choose Orleans

It is important to understand when not to choose a specific technology or framework. In this section, let's learn about when to choose and when not to choose Orleans.

Orleans is a great framework to build distributed applications. But if you have a single grain that takes 50% of all the requests, it is going to be a challenge to scale the system. If you have a request that needs to communicate with 50 other grains to serve the response, it is going to add latency and will have an impact on throughput. In both the scenarios mentioned above, the main problem is granularity. In all the Microsoft systems mentioned in the previous section that use Microsoft Orleans, the state is naturally partitionable and it nicely maps grains. It can be more easily distributed across a cluster of machines.

To summarize, think of the granularity while working with Orleans. Orleans may not be the best choice when data is shared across grains, when there's a small number of entities that are multi-threaded, when coordination is required with many other grains to serve a request, or in the case of operations that run for a long time.

Some scenarios where Orleans is a fit are the Internet of Things, social graphs, real-time analytics, mobile backend services, interactive entertainment, and gaming systems.

In this book, we are going to develop one such system, Distel, a distributed hotel management system. Distel supports a fictitious company that runs hotels in holiday destinations across the globe with thousands of partners and millions of patrons.

Summary

It is not simple and straightforward to design a distributed system. To achieve the ideal system, there are a lot of obstacles, such as concurrency, scalability, and fault tolerance, that must be overcome. Microsoft Orleans helps us by abstracting many of the challenges posed by distributed system development. It lets the developer focus on the core business logic. In this chapter, we have seen the challenges we face with an N-tier system and how Orleans assists developers in addressing them. We have also seen the key design choices made by Orleans, which we are going to put into practice in the upcoming chapters.

After a lot of theory and concepts in this chapter, it is time to get our hands dirty. In the next chapter, we will learn more about cloud-native grain objects and the systems that host them, silos.

Questions

1. What is the basic building block of an Orleans-based application?

    A. Actor

    B. Grain

    C. Seed

    D. None of the above

    Answer – B

2. When was the actor model originated?

    A. 1973

    B. 2000

    C. 2015

    D. The 19th century

    Answer – A

3. In Orleans, the process that hosts grains is referred to as _____.

    A. Forest

    B. Cluster

    C. Silo

    D. No specific name

    Answer – C

4. All operations in grains must return `Task` or `Task<T>` or `ValueTask` or `ValueTask<T>`.

    A. True

    B. False

    Answer – A

4

Understanding Grains and Silos

In the first part of this book, we learned about distributed applications and different patterns for implementing them. We also learned about the motivation behind building the Orleans framework and comprehended the major architectural decisions that went into designing the features of Orleans, as well as the rationale behind them.

In this chapter, we will delve deep into building a distributed application with Orleans. We will start this chapter by understanding the basic building blocks of Orleans, grains and silos, and see how easy it is to get an Orleans application up and running. This chapter will lay the foundation for building Orleans applications. We will also learn about the choices that developers can make to have better performance. Toward the end of this chapter, we will learn how to co-host an Orleans application with the ASP.NET Core application.

In this chapter, we will learn about the basic building blocks of Orleans:

- Say hello to Orleans – grains and hosting silos
- Creating your first Orleans application
- Understanding grain identity
- Stateless worker grains
- Request scheduling in Orleans
- Understanding silo configuration and co-hosting

Technical requirements

To follow along with the chapter, you need the following:

- Visual Studio 2022 Community Edition

- An understanding of building .NET applications

- The code used in this chapter can be found at `https://github.com/PacktPublishing/Distributed-.NET-with-Microsoft-Orleans/tree/main/Chapter04`.

Say hello to Orleans

In this section, we will build our first application using Orleans. Before we build the application, let's first understand more about grains and silos.

Cloud-native objects – grains

Grains are the main building blocks of applications built with Orleans. They are fundamental units of isolation, distribution, and persistence. They represent application entities. Grains are analogous to classes in the object-oriented programming model. Like classes encapsulate the data and the behavior of an entity, grains encapsulate the state and behavior of an application entity, as shown in *Figure 4.1*. Once a grain is activated, it will have a physical existence and an activated grain is called an **activation**. Activations are like objects in the object-oriented programming model. Activations can represent the digital twin of a real-world entity.

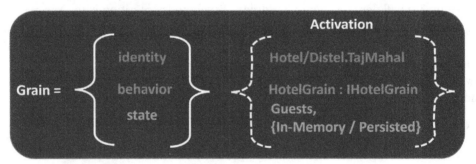

Figure 4.1 – Orleans grain

The preceding figure represents the hotel grain that we are going to build. It has the identity `Hotel/Distel.TajMahal` to uniquely identify the grain activation and implements the behavior defined in the `IHotelGrain` interface, for example, `CheckInGuest`. The hotel grain will have the state that is represented by the attributes, for example, hotel name, guests, location, and so on.

One of the challenges we saw with distributed applications in the previous chapter was the latency introduced due to data read and write operations with every request. To some extent, we can solve this by storing the hot data in the cache. With grains we can have the state loaded into the memory, so we do not need a separate cache. Upon activation, the grain will load the hot or warm data into the in-memory state and will be served from the state for the subsequent requests. The cold data can be fetched from the database when needed. Since the state of the grain is maintained in the in-memory state, it is the responsibility of the developer to write the in-memory state to persistent storage such as the database. The write message to the grain updates the state and the grain should update the database. Grains act like a smart write-through cache. They are scalable as they are spread across the cluster. We will learn more about grain state persistence in the next chapter.

In this section, we have learned about grains. In the next section, let's learn about another important concept of Orleans, which is the silo.

Hosting silos

Where are grains activated? In silos. Silo hosts can execute grains' behavior. You can visualize a silo as an Orleans runtime. Clients interact with grains hosted in silos through grain interfaces. Silos manage the life cycle of grains. Orleans may be deployed locally, on a single node for development and testing, or on a cluster of servers running in a data center or in the cloud. All the different deployment configurations are driven through the silo configuration. A cluster is like a big computer and grains are spread around a cluster.

Now that we have seen two key concepts in Orleans, grains and silos, let's build our first application in Orleans in the next section.

Creating your first Orleans application

It is time for us to create our first Orleans application. In this section, we are going to build a simple hotel grain, which will greet a guest. A typical Orleans solution structure contains a host application (it could be a console application, a web application, or a native application) that has the Orleans runtime and hosts grain activations, grain interfaces, and grain implementations.

Let's follow these steps to create our first Orleans solution:

Step 1: Create solution:

1. Create a new solution using the template **Blank Solution** in Visual Studio and name it `Distel`.

Step 2: Define grain interfaces: Grain interfaces define the different types of grains involved and their behavior:

1. Add a new project to this solution using the project template **Class Library**, name it `Distel.Grains.Interfaces`, and select the target framework **.NET6**.

2. To this project, add the following NuGet references:

```
Microsoft.Orleans.CodeGenerator.MSBuild
Microsoft.Orleans.Core.Abstractions
```

> **Important Note**
>
> `Microsoft.Orleans.CodeGenerator.MSBuild` is the code generation package that generates the code required for the custom serializer that we saw in the previous chapter. This package uses code generators to generate code. This package should be added to all those projects that contain grains, grain interfaces, and the types that are sent to and from grains.
>
> `Microsoft.Orleans.Core.Abstractions` is the library that defines the Orleans types that are needed for building Orleans applications. It defines types such as `IGrain`, `IGrainService`, and so on.
>
> `Microsoft.Extensions.Logging.Console` is not specific to Orleans. This is the package that provides the functionality required for console logging.
>
> `Microsoft.Orleans.Server` is a metapackage to build and start a silo. It includes the packages `Microsoft.Orleans.Core.Abstractions`, `Microsoft.Orleans.Core`, `Microsoft.Orleans.OrleansRuntime`, and `Microsoft.Orleans.OrleansProviders`.
>
> `Microsoft.Orleans.Client` is a metapackage to build an Orleans client. It includes the packages `Microsoft.Orleans.Core.Abstractions`, `Microsoft.Orleans.Core`, `Microsoft.Orleans.OrleansRuntime`, and `Microsoft.Orleans.OrleansProviders`.

3. Now that we have the required libraries added, add an interface, name it
    IHotelGrain, and define it as shown in the following code snippet:

```
public interface IHotelGrain :IGrainWithStringKey
{
      Task<string> WelcomeGreetingAsync(string
        guestName);
}
```

In the preceding code snippet, the IHotelGrain interface is inherited from the
interface IGrainWithStringKey. This tells Orleans that the identity of the hotel
grain is a string. We will learn more about grain identity later in this chapter. The
IHotelGrain interface defines the action WelcomeGreetingAsync, which the
hotel grain will support.

Step 3: Grain implementation: Here we define the details of the grain and implement the
behavior:

1. Add a new project to this solution using the project template **Class Library**, name it
    Distel.Grains, and select the target framework **.NET6**.

2. Add the following NuGet packages to this project:

```
Microsoft.Orleans.CodeGenerator.MSBuild
Microsoft.Orleans.Core.Abstractions
Microsoft.Extensions.Logging.Console
```

3. Add a project reference of Distel.Grains.Interfaces to this project.

4. Now add the HotelGrain class to this project and implement it as shown in the
    following snippet, which implements the IHotelGrain interface:

```
public class HotelGrain : Grain, IHotelGrain
{
        private readonly ILogger logger;
        public HotelGrain(ILogger<HotelGrain> logger)
        {
            this.logger = logger;
        }
        public Task<string>
          WelcomeGreetingAsync(string guestName)
        {
            logger.LogInformation($"\n
```

```
                    WelcomeGreetingAsync message
                    received: greeting = '{guestName}'");
               return Task.FromResult($"Dear {guestName},
               We welcome you to Distel and hope you
               enjoy a comfortable stay at our hotel.
               ");
        }
    }
```

The `WelcomeGreetingAsync` method implementation is basically generating a greeting message for the guest and returning it to the caller. If you notice, the grain method `WelcomeGreetingAsync` returns a `Task` object. Recall from the previous chapter that every operation in grain will return a `Task` object.

Step 4: Building the host: We'll build a silo to host the grains we defined in the previous step:

1. Add a new project to this solution using the project template **Console Application**, name it `Distel.Host`, and select the target framework **.NET6**.

2. Add the following NuGet packages to this project:

```
Microsoft.Extensions.Logging.Console
Microsoft.Orleans.Server
```

3. Add a project reference of `Distel.Grains` to this project.

4. To build the silo and start it, add the following code snippet to the `Program.cs` file's top-level statements. This code configures the host builder to host an Orleans silo and runs the application:

```
IHost host = Host.CreateDefaultBuilder(args)
    .UseOrleans(( builder) =>
    {
        builder.Configure<ClusterOptions>(options =>
        {
            options.ClusterId = "dev";
            options.ServiceId = "DistelService";
        })
        .UseLocalhostClustering(11111, 30000)
        .ConfigureApplicationParts(parts =>
          parts.AddApplicationPart
          (typeof(HotelGrain).Assembly)
```

```
            .WithReferences())
        .ConfigureLogging(logging =>
        logging.AddConsole());
    })
    .Build();
await host.RunAsync();

// Dispose the host
host.Dispose();
```

As we see in the preceding code snippet, `CreateDefaultBuilder` is used to create .NET Host Builder and the `UseOrleans` extension method is used in configuring the silo. The key aspects that we configure using silo are the following:

- **The clustering provider** – The services built in Orleans are deployed on a cluster of nodes. The clustering provider helps in the reliable management of membership. In the preceding code snippet, `UseLocalhostClustering` is configuring the clustering provider. The provider for storing the cluster information, `LocalhostClustering`, uses an in-memory provider. `LocalhostClustering` should not be used in production scenarios. We will learn more about different clustering providers in later parts of the book.

- **Clustering information** – The details of the cluster identity; it has two main fields:

  - `ClusterId` – The identity of the cluster. This can change with deployments. Normally, this is configured as the cloud region name, or some unique identifier driven through configuration.

  - `ServiceId` – The identifier of the service, this should not change with deployments. Normally, this is configured as the service name. `ServiceId` is used by some of the providers to identify the application, such as the persistence provider. `ClusterId` along with `ServiceId` is used to support blue-green deployment.

- **Application parts** – The Orleans runtime uses the configured application parts to scan grains, grain interfaces, and the serializer. If this is not configured, the Orleans runtime will scan all the assemblies, which will impact the system's startup performance. So, it is always recommended to configure application parts. In the preceding code snippet, we configured `typeof(HotelGrain).Assembly` to scan for grains. All the types sent to or from grains should be marked serializable and be included in application parts. The projects containing these types should include the Orleans code generators.

- **Endpoints** – Orleans has two endpoints configured, one is for silo-to-silo communication and the other is for client-to-silo communication. In the preceding code snippet, the endpoint configuration is abstracted inside `UseLocalhostClustering`. It configures `IPEndpoint` as `Loopback`, `siloPort` as `11111`, and `gatewayPort` as `30,000`. It gives an option to overwrite these configurations. A silo port is used for communication between silos within the same cluster. The gateway port is used for communication between clients and silos in the same cluster. Orleans provides more endpoint configurations, which we cover in the latter part of this book.

5. The host should be disposed of after it is stopped to ensure that the dependency injection container is disposed of. The code snippet shown in the previous step calls the `Dispose` method on the host to dispose of the resources.

Step 5: Build the client: We'll build a simple console client to interact with the Orleans service we built in the previous step:

1. Add a new project to this solution using the project template **Console application**, name it `ConsoleClientApp`, and select the target framework **.NET6**.

2. Add the following NuGet packages to this project:

```
Microsoft.Extensions.Logging.Console
Microsoft.Orleans.Client
```

3. Add a project reference of `Distel.Grains.Interfaces` to this project.

4. Add the following code to `Program.cs` top-level statements:

```
IClusterClient client = new ClientBuilder()
    .UseLocalhostClustering()
    .Configure<ClusterOptions>(options =>
    {
        options.ClusterId = "dev";
        options.ServiceId = "DistelService";
    })
    .ConfigureLogging(logging => logging.AddConsole())
    .Build();

await client.Connect();
```

This code snippet is using `ClientBuilder` to build the Orleans client to connect to a silo.

5. Add the `SendWelcomeGreeting` method to the `Program.cs` file as shown in the following code, which gets the hotel grain reference and gets the welcome greeting from the activated grain:

```
private static async Task
  SendWelcomeGreeting(IClusterClient client, string
  guest)
{
    var hotel =
      client.GetGrain<IHotelGrain>("Distel.Agra");
    var response = await
      hotel.WelcomeGreetingAsync(guest);
    Console.WriteLine($"\n\n{response}\n");
}
```

In the preceding code snippet, we are doing the following:

- We are calling the `GetGrain` method on the client with the parameter passed in as `Distel.Agra`, which is the grain identity.

- When the silo receives this request, it activates the grain with the identity as `Distel.Agra` and passes the proxy reference to the client.

- Now the client can call the grain proxy reference to send messages – in this case, `WelcomeGreetingAsync`.

- When the message is received by the grain activation, it processes it and sends back the response.

Have you noticed? It is as simple as calling a method defined in a class object. A grain reference is independent of the physical location of the grain. We can have multiple references for the same grain. All the actions we call on the grain reference are executed on the actual grain.

6. Now add the following code to the top-level statements to call the
    SendWelcomeGreeting method for a given guest name:

```
while (true)
{
    Console.Write("Please enter guest name. Type
        'exit' to close: ");
    var guest = Console.ReadLine();
    if (guest == "exit")
        break;
    await SendWelcomeGreeting(client, guest);
}
Console.WriteLine("Closing the client \n");
await client.Close();
client.Dispose();
```

In the preceding code snippet, for every guest arriving, we are calling
SendWelcomeGreeting iteratively to get the greeting from the hotel grain. Every
time we call GetGrain, we get the same grain activation. Here the client always
assumes that grain activation is always available.

The following screen clipping shows the solution structure of our application after
completing all the above-mentioned steps.

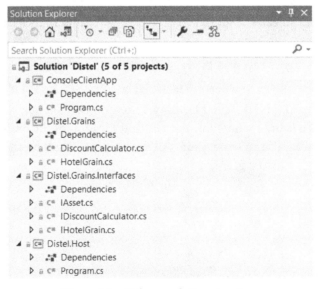

Figure 4.2 – Orleans solution structure

Step 6: Run the application:

1. Set both `Distel.Host` and `ConsoleClientApp` as startup projects by right-clicking on the solution and selecting **Set Startup Projects**.

2. Now run the solution by pressing *Ctrl + F5* or clicking **Start without debugging** from the **Debug** menu. You will see `Distel.Host` and `ConsoleClientApp` running in two different console windows.

3. Enter the name of the guest when prompted for the guest's name in the `ConsoleClient` app. You will see the response shown in the following screenshot.

Figure 4.3 – Orleans solution structure

Here `ConsoleClientApp` gets the hotel grain reference and activates the grain with the `Distel.Agra` identity. It then calls the `WelcomeGreetingAsync` method on it to get the welcome greeting message generated by the grain.

In this section, we have built a fully functional and incredibly simple Orleans application. While building our first application, we have seen different Orleans NuGet packages used in the Orleans application that provide the functionality. We have learned about some of the key silo configurations. We have also experienced how simple it is to get an Orleans application up and running. It all seemed magical. Let's understand what is happening behind the scenes and how the client call reaches the target grain.

What is happening behind the scenes?

In the last chapter, we learned that Orleans generates some code behind files to send and receive messages from the client to the sender. We also saw the library that is responsible for the code generation. Let's take a closer look at the code behind files generated for the application we built in the last section:

- The generated code file will be under `obj/Debug/{{respective .Net version}}` of the grain interfaces project. In our case, it will be under `obj/Debug/netstandard2.1`.

- The filename will be `{{Interface project name}}.orleans.g.cs`, that is, `Distel.Grains.Interfaces.orleans.g.cs`, as shown in the following screenshot.

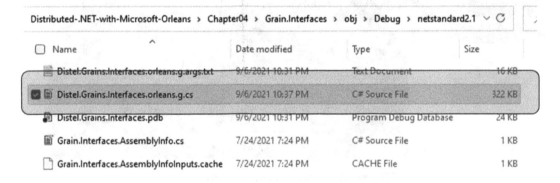

Figure 4.4 – Files generated by Orleans

If we open the file, we will notice two important classes,
`OrleansCodeGenHotelGrainReference` and
`OrleansCodeGenHotelGrainMethodInvoker`.

When we call `GetGrain` from the client as shown in the following code snippet,
Orleans gives us a reference to the generated `GrainReference` class, that is,
`OrleansCodeGenHotelGrainReference`:

```
var hotel = client.GetGrain<IHotelGrain>("Distel.Agra");
```

This reference class knows about the grain type and its identity. It does not know anything
about the location of the grain activation. When the `WelcomeGreeting` method is
called on the `GrainReference` class, the method call will be turned into a message to
the runtime by the `GrainReference` class by calling `InvokeMethodAsync`, shown in
the following figure.

```
internal class OrleansCodeGenHotelReference : global::Orleans.Runtime.GrainReference, global::Grain.Interfaces.IHotel
{
    // Code removed for brevity
    public override int InterfaceId => unchecked((int)0x8343A5B6);
    public override ushort InterfaceVersion => 0;
    public override string InterfaceName => "IHotel";
    public override bool IsCompatible(int interfaceId) => interfaceId == unchecked((int)0x8343A5B6);
    public override string GetMethodName(int interfaceId, int methodId)[...]

    global::System.Threading.Tasks.Task<string> global::Grain.Interfaces.IHotel.WelcomeGreeting(string guestName0)
    {
        return base.InvokeMethodAsync<string>((int)0x24E73AB8, new object[]{guestName0});
    }
}
```

Figure 4.5 – Generated GrainReference class

Now, the Orleans runtime transfers the message to the grain activation by finding the location of the grain activation by looking into the grain directory. Now to the activation. The message will be converted back to the method call by the runtime by calling the `Invoke` method from the `OrleansCodeGenHotelGrainMethodInvoker` class.

```
internal class OrleansCodeGenHotelMethodInvoker : global::Orleans.CodeGeneration.IGrainMethodInvoker
{
    public async global::System.Threading.Tasks.Task<object> Invoke(
        global::Orleans.Runtime.IAddressable grain,
        global::Orleans.CodeGeneration.InvokeMethodRequest request)
    {
        int interfaceId = request.InterfaceId;
        int methodId = request.MethodId;
        var arguments = request.Arguments;
        switch (interfaceId)
        {
            case unchecked((int)0x8343A5B6):
            {
                var casted = ((global::Grain.Interfaces.IHotel)grain);
                switch (methodId)
                {
                    case (int)0x24E73AB8:
                        return await casted.WelcomeGreeting((string)arguments[0]);
                    default:
                        ThrowMethodNotImplemented(interfaceId, methodId);
                        return null;
                }
            }

            default:
                ThrowInterfaceNotImplemented(interfaceId);
                return null;
        }
        // Code removed for brevity
    }
    public int InterfaceId => unchecked((int)0x8343A5B6);
    public ushort InterfaceVersion => 0;
}
```

Figure 4.6 – Generated GrainMethodInvoker class

Here is the flow diagram that gives the pictorial representation of different steps while invoking a method on a grain:

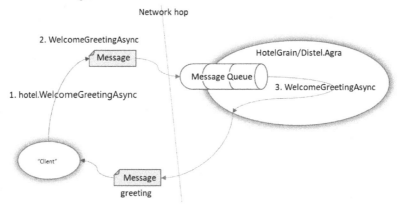

Figure 4.7 – Grain message flow

When the client makes the method call `WelcomeGreetingAsync` (step 1 in *Figure 4.7*) to the grain reference `OrleansCodeGenHotelGrainReference`, `OrleansCodeGenHotelGrainReference` converts the method call to a message (step 2 in *Figure 4.7*). The Orleans messaging system gets the reference to the grain activation with the key `Distel.Agra` by looking into the grain directory. The `WelcomeGreetingAsync` message will be enqueued to the `Distel.Agra` message queue. The enqueued messages will be processed by the activation in a **First In, First-Out (FIFO)** manner. The enqueued message will be converted to a method call by `OrleansCodeGenHotelGrainMethodInvoker`. Once the method `WelcomeGreetingAsync` is executed (step 3) by the grain, the response will be sent back to the caller in the form of a message.

If the activation is not found in the grain directory by the messaging system, the Orleans runtime activates the grain and places it somewhere in the cluster based on the placement constraints configured. We will learn more about placement constraints in *Chapter 8, Advanced Concepts in Orleans*.

Now we have got a fair idea of how a call from the client is served by a grain. In the next section, we'll learn about grain identity.

Naming a grain

When we create an object of a class using a new operator, what we get is the reference to the memory location. The variable identity is the name we define. If we create another object of the same class, we will get a reference to some other memory space.

As we learned before, grains represent some real-world entities. They exist virtually always. So, grains must have an identity to activate or interact with them. This is fundamentally the primary key of the grain. The scope of the key is the grain type. In the example that we saw in the previous section, `Distel.Agra` is the primary key of the hotel grain that we created. It represents the hotel with the name `Distel.Agra`. The scope of this identity is limited to hotel grains.

In Orleans, we have three main types of primary keys:

- Integer
- GUID
- String

In addition to these, there are compound keys:

- Integer + string
- GUID + string

The grain key is determined by the interface that we inherit the grain type from. Recall from the previous section that the grain interface IHotelGrain (refer to *Step 2: Define grain interfaces*), is inherited from IGrainWithStringKey. That means the key of the hotel grain is of type string. Similarly, to define the grain with the key as integer or GUID types, we use the IGrainWithIntegerKey and IGrainWithGuidKey interfaces respectively. To define compound keys, we use IGrainWithGuidCompoundKey for {GUID + String} and IGrainWithIntegerCompoundKey for {Integer + String}. For most scenarios, we will be using long, GUID, or string keys.

To understand the grain identity better, let's now do the following:

1. Add the GetKey method to the IHotelGrain grain interface and add implementation to HotelGrain as shown in the following code snippet:

```
public interface IHotelGrain : IGrainWithStringKey
{

    Task<string> WelcomeGreetingAsync(string
        guestName);
    Task<string> GetKey();
}
public class HotelGrain : Grain, IHotelGrain
{

......
public Task<string> GetKey()
{

    return
    Task.FromResult(this.GetPrimaryKeyString());

}

......

}
```

2. Now call the GetKey method on HotelGrain and print it along with
    IdentityString as shown in the following code snippet:

```
private static async Task
   SendWelComeGreeting(IClusterClient client, string
      guest)
{
    var hotel =
       client.GetGrain<IHotelGrain>("Distel.Agra");
    Console.WriteLine("Hotel Grain PrimaryKey : " +
       await hotel.GetKey());
    Console.WriteLine("Identity String : " +
       hotel.GetGrainIdentity().IdentityString);

    var response = await
       hotel.WelcomeGreetingAsync(guest);
    Console.WriteLine($"\n\n{response}\n\n");
}
```

3. Now, if we run the solution, we see the output that follows. Here, we see the key as
    the grain key we specified, that is, Distel.Agra, and the identity string formed
    by Orleans is *grn/Distel.Grains.HotelGrain/0+Distel.Agra-
    0xB34A441A. This is formed using the type of the object, that is, grain, grain key,
    and the hash code of the grain's identity.

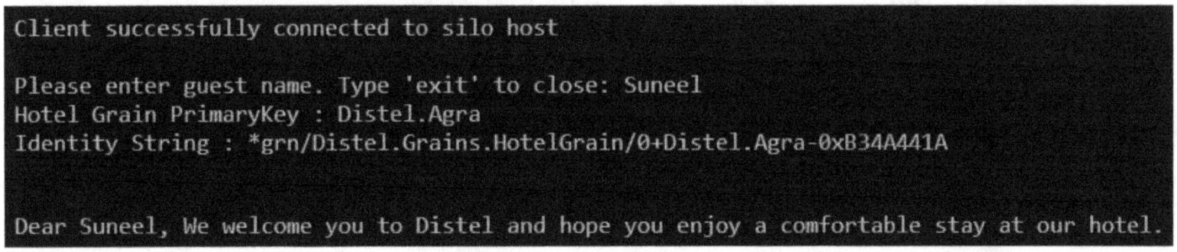

```
Client successfully connected to silo host

Please enter guest name. Type 'exit' to close: Suneel
Hotel Grain PrimaryKey : Distel.Agra
Identity String : *grn/Distel.Grains.HotelGrain/0+Distel.Agra-0xB34A441A


Dear Suneel, We welcome you to Distel and hope you enjoy a comfortable stay at our hotel.
```

Figure 4.8 – Sample output

Till now, we have talked only about grains that are tied to an entity such as `HotelGrain`. They possess some state, such as a list of customers onboard. These are called stateful grains. We will learn more about grain state in *Chapter 5*, *Persistence in Grains*. There are situations where we just need to perform some operations and the grain won't possess any state. In the next section, we will learn about stateless grains, which help to handle such scenarios.

Stateless worker grains

As we know, grains in Orleans will have only one activation at a given time to process messages received. This is the fundamental design principle behind Orleans grains. Consider a situation where we need to transform the data received from the client, we need to calculate a discount based on a price for our Distel application, or we need to route messages to another grain. In all these scenarios, we just need to process or route the data received. Those operations are not tied to a specific entity. To support such scenarios, Orleans has stateless worker grains.

To define a stateless worker grain, we just need to add the `[StatelessWorker]` attribute to the grain class. Now, let's build a simple discount computation grain for Distel by following these steps:

1. Add the grain interface `IDiscountCalculator` to the `Distel.Grains.Interfaces` project as shown in the following code snippet:

    ```
    public interface IDiscountCalculator :
      IGrainWithIntegerCompoundKey
    {
            Task<decimal> ComputeDiscount(decimal price);
    }
    ```

    The preceding code snippet defines `IDiscountComputeGrain` with `ComputeDiscount` functionality.

2. Implement the `DiscountCalculator` grain as shown in the following code snippet and add it to the `Distel.Grains` project:

    ```
    [StatelessWorker]
    public class DiscountCalculator : Grain,
      IDiscountCalculator
    {
        public Task<decimal> ComputeDiscount(decimal
          price)
    ```

```
{
    var discount = price switch
    {
        > 100 => price * 0.1M,
        > 50 => price * 0.05M,
        _ => 0
    };
    return Task.FromResult(discount);
}
}
```

In the preceding code snippet, we attributed `DiscountCalculator` with the `StatelessWorker` attribute. This indicates to the runtime that it is the stateless worker grain. The simple implementation of the `ComputeDiscount` function computes the discount as 10% if the price is more than $100 and 5% if the price is above $50, otherwise there's no discount.

3. Let's call this grain method from `ConsoleClientApp` as shown in the following code fragment:

```
var discountWorker =
    client.GetGrain<IDiscountCalculator>(0);
var discount = await
    discountWorker.ComputeDiscount(150);
Console.WriteLine($"Discount for the Amount ${150} is
    ${discount}");
```

4. Now run the application. You see will the output as follows. The discount for the billed amount of $150 is calculated and printed as $15:

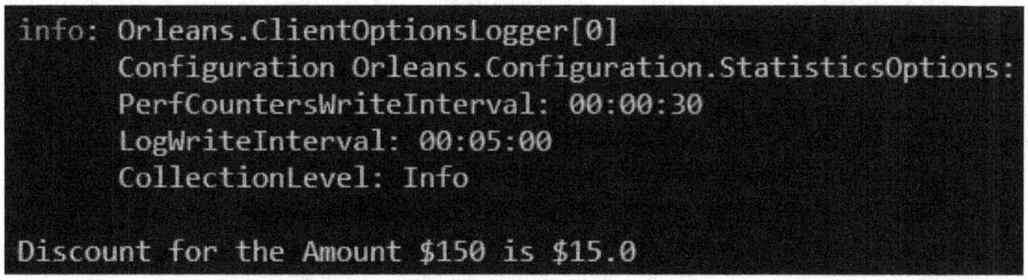

```
info: Orleans.ClientOptionsLogger[0]
      Configuration Orleans.Configuration.StatisticsOptions:
      PerfCountersWriteInterval: 00:00:30
      LogWriteInterval: 00:05:00
      CollectionLevel: Info

Discount for the Amount $150 is $15.0
```

Figure 4.9 – Sample output of the DiscountCalculator grain

The name says stateless worker. It does not mean that the grain should not have state. The grain can load and keep the in-memory state required for performing the operations. For example, it can load the rules to compute the discount and keep that in memory.

The properties of stateless worker grains are as follows:

- Stateless worker grains are activated on the same silo if the silo has an implementation of the requested stateless grain. That means they will be activated on the same silo as where the caller is. So, they will not incur any network costs. If the silo does not have the implementation of the requested stateless grain, it will be activated on a random compatible silo.

- Stateless worker grains will have multiple activations on different silos.

- If the existing stateless worker grains are busy, the Orleans runtime will create additional activations. The maximum number of activations is limited by the number of CPU cores unless the developer explicitly configures it using the `maxLocalWorkers` configuration as shown in the following code snippet:

```
[StatelessWorker(maxLocalWorkers : 1)]
public class DiscountCalculator : Grain,
    IDiscountCalculator
```

  As per the preceding code snippet, there can be only one activation for the `DiscountCalculator` grain.

- Because of the second and third points, two subsequent requests to a stateless worker grain may be processed by two different activations.

We have learned how stateless worker grains are used in scenarios to perform stateless operations that are not tied to a specific entity. Moreover, our understanding is that a grain processes one request at a time in a single thread. But there are circumstances where it is desirable for a grain activation to perform other operations while the current request is waiting for an asynchronous operation to complete. In the next section, we will learn about the flexibility provided by Orleans to manage such scenarios.

Request scheduling in Orleans

In the previous chapter, we learned that one of the design decisions of Orleans is to use cooperative multi-tasking. By default, the Orleans runtime will execute incoming requests on a given grain from start to completion before processing the next request. Orleans gives some control to developers where the current request is for an asynchronous operation (for example, an I/O operation) to complete, to let other requests be processed.

Let's understand the default request scheduling behavior in Orleans with an example. Consider the following code snippet of the `HotelGrain` implementation to handle the scenario of a guest transferring from one hotel to another hotel in our chain. At the time of check-in to the new hotel, we need to compute the amount due from the previous hotel:

```
public class HotelGrain : Grain, IHotelGrain
{
    public Task<decimal> ComputeDue(string guestName)
    {
        return Task.FromResult(100.00M);
    }

    public async Task OnboardFromOtherHotel(IhotelGrain
        fromHotel, string guestName)
    {
        logger.LogInformation($"Onboarding the guest from
            other hotel {guestName}");
        await fromHotel.ComputeDue(guestName);
        logger.LogInformation("2");
    }
}
```

When a guest is transferring from one hotel to another hotel in our chain, `OnboardFromOtherHotel` will get called. In `OnboardFromOtherHotel`, we will first fetch the due from the source hotel by calling the `ComputeDue` function and then onboard the guest in the hotel. The following code snippet shows how different grain invocations happen in this scenario:

```
var tajMahal =
    grainFactory.GetGrain<IHotelGrain>("TajMahal");
var charminar =
    grainFactory.GetGrain<IHotelGrain>("Charminar");
await tajMahal.OnboardFromOtherHotel(charminar, "Shyam");
```

For this sequence of code lines, the code flow will be something like this:

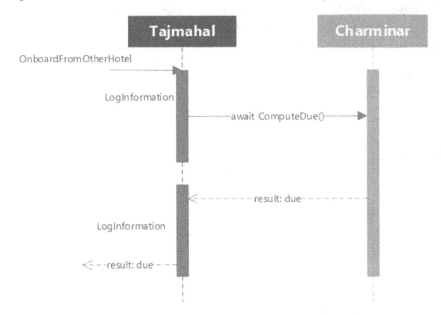

Figure 4.10 – Control flow during the grain call

- The call arrives at the grain TajMahal and logs the information.
- The Tajmahal grain issues the call to the Charminar hotel grain to get the due amount.
- The Charminar grain computes the due and returns it to the caller grain Tajmahal.
- The Tajmahal grain receives the computed due amount and logs the information.

While the Tajmahal grain is awaiting the call to the Charminar grain to compute the amount due, as per the default behavior of the Orleans runtime, it won't process any other requests. So, if both grains were to call each other simultaneously, as you might have guessed, there would be a possibility of deadlock.

Consider the following code for a hypothetical scenario where a guest is transferring from Tajmahal to Charminar and another guest from Charminar to Tajmahal:

```
await Task.WhenAll(
        tajMahal.OnboardFromOtherHotel(charminar,
        "Shyam"),
        charminar.OnboardFromOtherHotel(tajMahal, "Ram")
);
```

In the preceding code, if it happens that the `TajMahal` grain is waiting for the `Charminar` grain to compute the due and the `Charminar` grain is waiting for the `TajMahal` grain to compute the due, then there will be a deadlock as per the default implementation of Orleans request scheduling. In such a situation, the runtime will throw a timeout exception, which will be bubbled up to the caller.

To handle such situations, Orleans gives some control to the developer to execute multiple requests in an interleaving way. This behavior is called reentrancy. In the next section, let's learn about reentrant grains.

Interleaving grain code execution

While the default request execution behavior of Orleans will protect the developer from complications that arise with concurrent modification of state by different requests, it adds the cost, as we saw in the previous section, that it can lead to deadlocks if grain calls in a system can form cycles. Waiting for the request to complete before processing the next request might impact the performance. For example, while the grain is waiting for an I/O operation or an asynchronous web service call, it won't be performant to stop processing other requests to the grain.

Orleans provides developers with different options to allow a request to execute in an interleaving way. The options provided by Orleans are as follows:

- Reentrant grain
- Interleaving method execution
- Interleaving method execution decided by a predicate

Reentrant grain

Reentrant grains may process other requests while the current request is awaiting an asynchronous operation to complete. Please note that all the execution for a given grain activation will happen in a single-thread manner. So, two different pieces of grain code will never be executed parallelly. We can mark any grain as reentrant by marking the grain implementation class with the `[Reentrant]` attribute. The following code snippet shows a reentrant grain implementation:

```
[Reentrant]
public class VehicleGrain : Grain, IVehicle
{
}
```

Interleaving method execution

In situations where only a few actions have interleaving behavior in a grain, we mark those functions with the `[AlwaysInterleave]` attribute. This will notify the runtime to execute these methods in an interleaving way. In the following code snippet, the `GetTimeToService` method of the `IAsset` grain will always be executed in an interleaving way as we marked it with the `[AlwaysInterleave]` attribute:

```
public interface IAsset : IGrainWithStringKey
{
    [AlwaysInterleave]
    Task<TimeSpan> GetTimeToService();
}
```

Interleaving method execution decided by a predicate

There can be situations where we want to have the interleaving behavior conditionally. For this, Orleans supports reentrancy using a predicate. To do this, we will define an attribute to determine the interleaving behavior of the current request under execution.

We'll leave the implementation of this as an exercise for you. You may want to follow the Orleans documentation:

https://dotnet.github.io/orleans/docs/grains/reentrancy.html#reentrancy-using-a-predicate

Reentrancy will give better performance results, but we should think about the state change between interleaved calls, which leads to concurrency bugs. With a thoughtfully crafted flow, we can get better performance with reentrancy.

This section showed us how to take advantage of interleaved calls with reentrancy. In the next section, we will learn about the most popular hosting technique used with ASP.NET Core applications.

Co-hosting silos with ASP.NET core

Typically, an Orleans application is intended to run the core business functionality. Handling cross-cutting concerns, authentication, authorization, request throttling, and so on is not the primary responsibility of Orleans. So, it is recommended not to have direct interactions between external clients and Orleans silos. There should be a gateway before the Orleans application that routes the user requests. The Orleans application should not directly take the user requests.

Figure 4.11 – Orleans N-tier application

Consider a traditional N-tier architecture as shown in the preceding figure. Each tier is physically separated and will mostly be running on different machines. Here, the service tier will be the interfacing layer to the business layer built in Orleans. The service layer takes the user requests coming from the presentation tier and processes them by interacting with the business tier built in Orleans. The service tier and the business tier will be running on different processes on the same machine or different machines.

One of the most interesting features introduced with Orleans 3.0 is the ability to co-host the Orleans application with an ASP.NET Core application. With this, we can run both Orleans and an ASP.NET Core application in the same process and they both share the service registrations. What this means is that they both share the same service provider, logging, and so on configured in the host builder. Co-hosting helps in routing the requests and avoids the need for a gateway. In the case of `StatelessWokrer` and `PreferLocal`, the requests can be served locally, which will improve latency and throughput.

Another important benefit that co-hosting enables is being able to include the status of the Orleans silo in the ASP.NET Core health check endpoint. If we are running the service on Azure Service Fabric, Kubernetes, or AWS **ECS** (**Elastic Container Service**), this will help to determine the health of the Orleans cluster.

Let's go ahead and create a co-hosted application by following these steps:

1. Add a new project to the Distel solution using the ASP.NET Core Web API template and name it `Distel.WebHost`.

    > **Note**
    >
    > Select the **Enable Docker** option while creating the project. Though this is not required to co-host the application, we will be running our application in Docker in later chapters.

2. As we learned earlier in this chapter, add the `Microsoft.Orleans.Server` NuGet package reference to host the Orleans silo with the ASP.NET Core application.

3. Add Orleans to `IHostBuilder` by adding the following code snippet to the `Program.cs` file top-level statements:

```
builder.Services.AddEndpointsApiExplorer();
builder.Services.AddSwaggerGen();

builder.Host.UseOrleans(siloBuilder =>
{
    siloBuilder
    .UseLocalhostClustering()
    .Configure<ClusterOptions>(opts =>
    {
        opts.ClusterId = "dev";
        opts.ServiceId = "DistelAPI";
    })
    .Configure<EndpointOptions>(opts =>
    {
        opts.AdvertisedIPAddress =
            IPAddress.Loopback;
    });
});

var app = builder.Build();
```

Here, we use `UseOrleans` to configure the host builder to host our Orleans application. You may notice that the rest is as we configured the console host application earlier in this chapter. We named our cluster and service as `dev` and `DistelService` respectively. We are also using `UseLocalhostClustering` to configure the cluster and the Orleans application is available on the localhost loopback IP address.

4. Add an API controller to the `Distel.WebHost` project and name it `DistelController`. This controller will be responsible for all the requests related to Distel.

5. Add a GET API to fetch the greeting message for the guest at the time of guest check-in as shown in the following code snippet:

```
public class DistelController : ControllerBase
{
    private readonly ILogger<DistelController>
      _logger;
    private readonly IClusterClient clusterClient;

    public DistelController(ILogger<DistelController>
      logger, IClusterClient clusterClient)
    {
        this._logger = logger;
        this.clusterClient = clusterClient;
    }
    [HttpGet("welcome/{hotel}/{guestname}")]
    public async Task<IActionResult>
    WelcomeGuest([FromRoute]string hotel, [FromRoute]
      string guestname)
    {
        var hotelGrain =
          this.clusterClient.GetGrain
          <IHotelGrain>(hotel);
        var greeting = await
          hotelGrain.WelcomeGreetingAsync(guestname);
        return Ok(greeting);
    }
}
```

In the WelcomeGuest method, we are fetching a reference to the activation of the hotel grain with the key as the hotel name passed in the request. Then we call the WelcomeGreetingAsync method on the grain action to fetch the greeting by passing in the guest's name. The greeting message is then passed back to the caller. The IClusterClient interface used to get the grain activation is injected via constructor injection into the DistelController class. This is possible because the Orleans runtime and ASP.NET Core application are using the same service provider, as we discussed earlier in this section.

6. Now when you run the application, it will launch Swagger UI with the URL `https://localhost:5001/swagger/index.html`.

    Swagger UI will have the list of action methods exposed. Now try the `WelcomeGuest` API from Swagger UI. We see the output shown in the following figure.

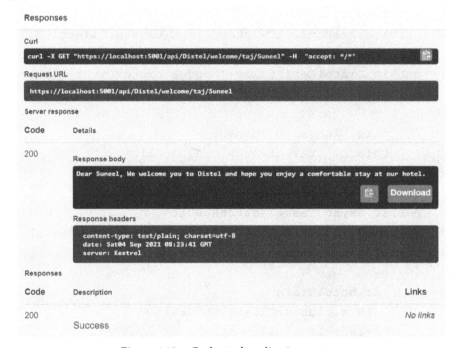

Figure 4.12 – Co-hosted application output

With this, we co-hosted Orleans and ASP.NET core applications in the same process. While building our Orleans application, we used `UseLocalhostClustering`, which should only be used for development purposes as it uses in-memory cluster management. It is not recommended for production scenarios. In the upcoming chapters, we will learn about different components and libraries available to host the silo.

Though for simplicity of understanding we took the example of a typical N-tier application, Orleans can be used in any of the cloud architecture styles that we discussed in *Chapter 2, Cloud Architecture and Patterns for Distributed Applications*, such as event-driven architecture or web-queue-worker.

Summary

In this chapter, we have experienced how easy it is to get a fully functional Orleans application up and running. We also learned about silo configurations and different types of grain identities. In addition to this, we have seen stateless worker grains along with scenarios to use them in. We also learned about the flexibility that Orleans provides with reentrant grains to optimize performance. Toward the end, we built an Orleans application co-hosted with an ASP.NET Core Web application, which enabled us to share the registered services as they both share the same host builder.

You may be wondering, if none of the grains we created have any state in them, then why did we talk about Orleans grain state management? In the next chapter, let's learn about grain state persistence.

Questions

1. What is a grain in Microsoft Orleans?

   A. Identity + Behavior + State

   B. Identity

   C. Identity + State

   D. Behavior + State

   Answer – A

2. The grain interface should be inherited from the _____ interface to have an integer key.

   A. IGrainWithStringKey

   B. IGrainWithGuidKey

   C. IGrainWithIntegerCompoundKey

   D. IGrainWithIntegerKey

   Answer – D

3. In Orleans, all the grain methods should support the async pattern.

   A. True

   B. False

   Answer – A

4. As the name says, the stateless worker grain cannot have state.

    A. True

    B. False

    Answer – B

5
Persistence in Grains

In the last chapter, we learned about Orleans grains and silos, which host grains. We also learned about one of the most prominent hosting patterns, co-hosting Orleans applications with an ASP.NET Core application. The grains we built in the last chapter do not have any state associated with them. In the real world, most applications will deal with data. This data (or state as we call it) should be stored in some persistent storage and be presented to the client application when needed.

In this chapter, we will learn how we manage state in Orleans, and in due course, we will cover the following topics:

- Understanding grain state persistence
- Creating a custom state provider
- Grains with multiple states
- Grains directly interacting with a database

Technical requirements

To follow along with this chapter, you will need the following:

- Visual Studio 2022 Community Edition
- An understanding of building .NET applications
- The Cosmos DB Emulator, from `https://aka.ms/cosmosdb-emulator`
- An Azure subscription

The code used in this chapter can be found at `https://github.com/PacktPublishing/Distributed-.NET-with-Microsoft-Orleans/tree/main/Chapter05`.

Understanding grain state persistence

As we learned in earlier chapters, grains can correspond to real-world entities. For example, in our Distel application, `HotelGrain` represents the digital life of a hotel in our chain of hotels. `UserGrain` represents the user or customer of our application. These grains will process the messages received from clients. `UserGrain` should hold user details such as name and address, which we call state. There is a method to update state. If `UserGrain` receives a message to update the address, `UserGrain` will update the address in the in-memory local state so that it can send the updated address when requested later. What happens when the grain deactivates? The internal in-memory state is lost. The application won't be able to serve the updated address. So, it is important to persist the grain state in a permanent database such as SQL Server or Azure CosmosDB. With the state persisted in the permanent storage medium, the grain can load the state from persistence storage during activation and update it as desired.

Orleans supports a number of persistent stores. They are supported through a plugin model. It supports Azure CosmosDB, Azure Table Storage, and Amazon DynamoDB. It also supports various relational databases, such as SQL Server and Oracle via the ADO. NET grain storage provider.

To understand how state persistence works, let's add functionality to get the current checked-in guests from the `HotelGrain`. During check-in, we will persist the guest details along with the period of the stay. During checkout, we will remove the user from the current guest list. We need to persist the state of checked-in guests to permanent storage so that we can retrieve it across multiple grain activations.

Adding grain state persistence using Azure Cosmos DB

Let's use Azure Cosmos DB as persistent storage in our sample application. Follow these steps to build this functionality:

1. **Reference persistence provider package**: Add the NuGet package reference to `Orleans.Persistence.CosmosDB` to our `Distel.WebHost` project.

> **Note**
>
> The `Orleans.Persistence.CosmosDB` package is provided by Orleans Community Contributions: `https://github.com/OrleansContrib`. This community is very active and has contributed many useful libraries and functionalities to Orleans.

2. **Update Silo configuration**: Add Cosmos DB storage to the silo builder in the
    `Program.cs` file as shown in the following code snippet:

```
builder.Host.UseOrleans(siloBuilder =>
{
    siloBuilder.UseLocalhostClustering()
        .AddCosmosDBGrainStorageAsDefault(opt =>
        {
            opt.AccountEndpoint = "<<Cosmos DB accoutn
                Endpoint>>";
            opt.AccountKey = "<<Cosmos DB account Key>>";
            opt.DB = "<<Cosmos DB Database name>>";
            opt.CanCreateResources = true;
        })
<<Code removed for brevity>>
});
```

In the preceding code snippet, we called
`AddCosmosDBGrainStorageAsDefault` to register the Cosmos DB provider.
`AccountEndpoint` represents the Cosmos DB endpoint. `AccountKey`
represents the secret key to access Cosmos DB. The DB configuration name is
set to the collection name of the Cosmos DB where the grain state is persisted.
Set `CanCreateResources` to true, if you want the Cosmos DB collection
configured with the DB parameter to be created automatically when the application
starts. `AddCosmosDBGrainStorageAsDefault` makes the configured Cosmos
DB resource as the default state provider for all the grains.

> **Note**
>
> In the preceding code, we have the Cosmos DB connection secret
> configured in plain text and assigned to `AccountKey`. This is done only for
> demonstration purposes. In production scenarios, we should be fetching such
> secrets from secret providers such as Azure Key Vault.

3. **State data model**: Let's now define the data model of the `UserCheckIn` state. Add a class to `Distel.Grains.Interfaces` and name it `UserCheckIn`. Define the `UserCheckIn` class as shown in the following snippet:

```
public class UserCheckIn
{
    public string UserId { get; set; }
    public DateTime CheckInDate { get; set; }
}
```

As defined in the model, the `UserCheckIn` state will store the `UserId` along with `CheckInDate`.

4. Let's define the grain action method to the grain interface `IHotelGrain` as shown in the following code snippet:

```
public interface IHotelGrain : IGrainWithStringKey
{
    Task<string> WelcomeGreetingAsync(string
      guestName);
    Task<string> GetKey();
    Task OnboardFromOtherHotel(IHotelGrain fromHotel,
      string guestName);
    Task<decimal> ComputeDue(string guestName);
    Task<string> CheckInGuest(UserCheckIn
      userCheckIn);
}
```

The `CheckInGuest` method takes the user's check-in details defined in the model `UserCheckIn`. We will be implementing this method in `HotelGrain` in the next step.

5. **Persist the state**: To make the grain use the state from the persistent storage provider, add the `checkedInGuests` state parameter to the `HotelGrain` constructor as shown in the following code snippet and add a using statement for the `Orleans.Runtime` namespace:

```
private IPersistentState<List<UserCheckIn>>
  checkedInGuests;
public HotelGrain(ILogger<HotelGrain> logger,
  [PersistentState("checkedInGuests")]
IPersistentState<List<UserCheckIn>> checkedInGuests)
{
    this.logger = logger;
    this.checkedInGuests = checkedInGuests;
}
```

In the preceding code, we added a constructor parameter, `checkedInGuests`, which is of the type `IPersistentState`. In the grain activation process, the `Grain` object is created and a reference to `IPersistentState` is injected into the constructor. The persistence state name will be determined by the `PersistentState` attribute, configured for the `checkedInGuests` state argument.

After creating the `Grain` object, the Orleans runtime loads the state from the configured persistent database and then calls the `OnActivateAsync` method. Developers can optionally override this method to perform any additional initialization tasks. Similar to `OnActivateAsync`, there is `OnDeactivateAsync`, which will be called by the runtime during grain deactivation. We can override this to perform any cleanup activities. The call to `OnDeactivateAsync` is not guaranteed in all cases. For example, there won't be a call to `OnDeactivateAsync` in the event of server failure.

6. Add the implementation of the `CheckInGuest` method to the `HotelGrain` class as shown in the following code:

```
public async Task<string> CheckInGuest(UserCheckIn
  userCheckIn)
{
    checkedInGuests.State.Add(userCheckIn);
    await this.checkedInGuests.WriteStateAsync();
    return "";
}
```

In the preceding code snippet, we are adding the checked-in users to the checkedInGuests state. The call WriteStateAsync on the checkiedInGuests state persists the state to the storage. In this case, we are writing the state to the permanent storage with every change, that is, with every guest check-in. Developers can decide when to call WriteStateAsync based on the need as this will add latency. We may choose to persist state upon deactivation. For example, in a banking system, every transaction should be persisted, so we may go with saving the state with every transaction. In situations where it is acceptable to lose state partially, we may persist the data periodically. The cadence pattern that will be discussed in *Chapter 9, Design Patterns in Orleans*, can be used to persist state periodically. We will lose in-memory state changes, which are not persisted in the event of cluster failures. We should be considerate of this fact while deciding when to call WriteStateAsync.

7. Now add the controller action check to the DistelController controller, which will invoke the CheckInGuest method on HotelGrain as shown in the following code snippet:

```
[HttpPost("checkin/{hotel}/")]
public async Task<IActionResult> CheckIn([FromRoute]
    string hotel, [FromBody] UserCheckIn
    userCheckInDetails)
{
    var hotelGrain =
        this.clusterClient.GetGrain<IHotelGrain>(hotel);
    var alottedRoom = await
        hotelGrain.CheckInGuest(userCheckInDetails);
    return Ok(alottedRoom);
}
```

Here we are retrieving the HotelGrain activation for the given identity and calling the CheckInGuest method by passing the relevant data.

8. Now run the application and call the action we implemented in the previous step from Swagger UI.

9. Open the Cosmos DB Data Explorer to view the persisted data. You will see data like the following JSON. Please note a few fields have been removed for brevity:

```
{
    "id": "DistelAPI__GrainReference=00000000000000000
        00000000000000006fffffffa8d0b7cd+Charminar",
    "GrainType": "Distel.Grains.HotelGrain,
        Distel.Grains.checkedInGuests",
    "State": [
        {
            "UserId": "bhupesh.guptha",
            "CheckInDate": "2021-09-19T13:21:26.515Z",
        }
    ],
    "_etag": "\"00000000-0000-0000-ad59-
        9f343a0001d7\"",
    "PartitionKey": "Distel.Grains.HotelGrain,
        Distel.Grains.checkedInGuests",
    "_ts": 1632057846
}
```

If you look at the JSON, the `"id"` field has the `HotelGrain` name `Charminar`. `GrainType` and `PartitionKey` are populated with the grain type name (`HotelGrain`) and the persistent state name (`checkedInGuests`). Finally, the `"State"` field has the array of the checked-in guests' details.

In this section, we built persistence storage for `HotelGrain` using Cosmos DB. We have seen how the persisted state is made available upon activation of a grain. In the next section, let's build a custom storage provider.

Custom persistence provider

We learned earlier in this chapter that Orleans persistent storage providers are available for most of the traditional storage systems out there currently. What if there is no storage provider by Orleans or the community for the database system you use in your organization? Orleans is built on a plugin model; we can easily create a custom storage provider to support our own storage system.

In this section, we will implement a simple persistence provider for our Distel application using Azure file shares.

> **Note**
>
> Azure Files is not a traditional storage provider for persisting data. Its main use case is to provide file sharing via different protocols such as SMB, NFS, and FileREST. It can be mounted on workstations running Windows, macOS, and Linux. You can learn more about Azure Files here: `https://docs.microsoft.com/en-us/azure/storage/files/`.

Follow the steps described here to create an Azure Files storage provider:

1. Add a new project to our Distel solution with a *Class Library* project template and name it `Distel.OrleansProviders`.

2. Add a reference to the below-mentioned NuGet packages:

    - `Microsoft.Orleans.Runtime.Abstractions` – Offers the abstractions used in creating custom providers

    - `Azure.Storage.Files.Shares` – Offers the functionality required to interact with Azure Files

    Let's add a class with the name `AzureFileGrainStorage` to this project and inherit the `IGrainStorage` interface.

    The `IGrainStorage` interface defines the methods required to support storage providers.

```
namespace Orleans.Storage
{
    public interface IGrainStorage
    {
        Task ClearStateAsync(string grainType, GrainReference grainReference, IGrainState grainState);
        Task ReadStateAsync(string grainType, GrainReference grainReference, IGrainState grainState);
        Task WriteStateAsync(string grainType, GrainReference grainReference, IGrainState grainState);
    }
}
```

Figure 5.1 – IGrainStorage interface definition

The `ReadStateAsync` method is called by the runtime while loading the `Grain` state. `WriteStateAsync` is called to persist the `Grain` state to the storage provider. `ClearStateAsync` is called to clean up the state. Let's implement each of these methods in the coming steps.

Storage providers in Orleans should implement the `IGrainStorage` interface.

3. We will build our storage provider such that it persists the JSON serialized grain state in the files. We will name the files with the Grain identity. These files will be put inside a directory. The name of the directory will be grain type. This directory will be put inside the file share with a name configured by the developer while registering the storage provider.

The following diagram represents the Azure file storage structure for our Distel application. The Azure File storage provider will have the file share named distelshare and grain types HotelGrain and UserGrain are represented as directories. Each grain activation will have the state persisted in the file under the respective directories with the filename as the grain identity.

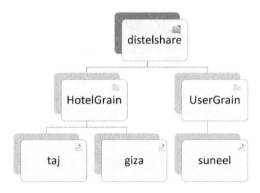

Figure 5.2 – Azure Files grain storage representation

The storage provider should be initialized at the time of silo initialization with the required resources, such as ShareClient in the case of AzureFileGrainStorage. Orleans provides lifecycle hooks to perform such operations. To participate in silo lifecycle events, the class should implement the ILifecycleParticipant<ISiloLifecycle> interface:

```
public class AzureFileGrainStorage : IGrainStorage,
    ILifecycleParticipant<ISiloLifecycle>
{
<<Code removed for brevity>>
    public void Participate(ISiloLifecycle lifecycle)
    {
    lifecycle.Subscribe(OptionFormattingUtilities.Name
        <AzureFileGrainStorage>(this._name),
        this._options.InitStage, this.Init, this.Close);
    }
}
```

In the preceding code snippet, the `AzureFileGrainStorage` class implements the `Participate` method from the `ILifecycleParticipant<ISiloLifecycle>` interface. The `Participate` method subscribes to the silo lifecycle events `Init` and `Close`. We will be executing the initialization tasks upon receiving the `Init` silo event and any cleanup tasks on the `Close` event.

4. To initialize `ShareClient`, we should be establishing the connection to Azure Files. For this, we need the share name and the connection string. These will be configured and injected during silo configuration. Define the `AzureFileStorageOptions` class as shown in the following code snippet:

```
public class AzureFileStorageOptions
{
internal const string ORLEANS_STORAGE_SHARE =
    "orleansstorage";
        public string ConnectionString { get; set; }
        public string Share { get; set; } =
            ORLEANS_STORAGE_SHARE;
        public int InitStage { get; set; } =
            DEFAULT_INIT_STAGE;
        public const int DEFAULT_INIT_STAGE =
            ServiceLifecycleStage.ApplicationServices;
}
```

This defines the default share name as `orleansstorage`, which can be overridden at the time of silo configuration.

5. Now let's implement the `Init` method, which initializes the storage client as shown in the following code snippet:

```
public async Task Init(CancellationToken ct)
{
<<Code removed for brevity>>
    try
    {
        this._shareClient = new
            ShareClient(this._options.ConnectionString,
            this._options.Share.ToLower());
        await this._shareClient.CreateIfNotExistsAsync();
```

```
            }
        catch
        {
//Log exception
            throw;
        }
    }
```

In the preceding code snippet, we will be instantiating the `ShareClient` instance using the connection string configured. We create the Azure file share if it does not exist already.

> **Note**
>
> `ShareClient` helps us to manipulate Azure Files. More about `ShareClient` can be found here: `https://docs.microsoft.com/en-us/dotnet/api/azure.storage.files.shares.shareclient`.

6. Let's implement the `WriteStateAsync` method as shown in the following snippet. The `WriteStateAsync` method will be called from the grain to persist the state. To write the state to the storage, we will first check for the existence of the state. If a persisted state exists, it will be overwritten after validating the version of the state by comparing the ETag. If there is no persisted state, then a new file will be created with the state object:

```
public async Task WriteStateAsync(string grainType,
    GrainReference grainReference, IgrainState
    grainState)
{
    string id = this.GetFileName(grainReference);
    var directory =
        this._shareClient.GetDirectoryClient(grainType);
    await directory.CreateIfNotExistsAsync();
    ShareFileClient file =
        directory.GetFileClient(id);
    try
    {
        await WritetoFile(grainState, file);
        grainState.RecordExists = true;
```

```
        }
    catch (Exception exc)
    {
        // Log Error
        throw;
    }
}
```

In the preceding code snippet, we are checking for the existence of the directory with the name of the grain type and then checking for the existence of the file with the grain ID as the name. If the file is present, then we compare the ETag of the incoming state with the storage provider ETag. If the ETag matches, we will overwrite the state on the provider with the incoming state by calling `WriteToFileAsync`. `WriteToFileAsync` writes the grain state to Azure Files and returns the updated ETag. For the implementation of `WriteToFileAsync`, please refer to the sample code of this book on GitHub. If the ETag does not match, that means the state is not consistent. We will throw the `InconsistentStateException` exception to the caller so that the caller can refresh the state and retry the write operation.

7. Let's now implement the `ReadStateAsync` method. `ReadStateAsync` will be called at the time of grain activation to load the state or when explicitly called, `ReadStateAsync` will read the state. Add the `ReadStateAsync` method to the `AzureFileGrainStorage` class as shown in the following code snippet:

```
public async Task ReadStateAsync(string grainType,
    GrainReference grainReference, IgrainState
    grainState)
{
    string id = this.GetFileName(grainReference);
    try
    {
        var directory =
        this._shareClient.GetDirectoryClient(grainType);
        if (await directory.ExistsAsync())
        {
            var file = directory.GetFileClient(id);
```

```
        if (await file.ExistsAsync())
        {
            ShareFileDownloadInfo download = await
              file.DownloadAsync();
            using (StreamReader reader = new
              StreamReader(download.Content))
            using (JsonTextReader jsonReader = new
              JsonTextReader(reader))
            {
                JsonSerializer ser = new
                  JsonSerializer();
                grainState.State =
                  ser.Deserialize(jsonReader,
                  grainState.State.GetType());
            }
            grainState.RecordExists = true;
            grainState.ETag =
              download.Details.ETag.ToString();
        }
    }
    if (grainState.State == null)
    {
        grainState.State =
          Activator.CreateInstance
          (grainState.State.GetType());
        grainState.RecordExists = true;
    }
}
catch (Exception exc)
{
    throw;
}
}
```

In the preceding code snippet, we are checking for the existence of the directory with the grain type as the name. If present, then we will check for the existence of the file with the grain ID as the name. If the file is present, we read the file content using the `ShareClient` object that we created in the `Init` method and load the state by deserializing the file content to the object type of the grain state. If the file or the directory is not present, then we will initialize the grain state to the initial state by creating the instance of the grain state type as the grain is being activated for the first time.

> **Note**
>
> The Orleans state class should have the default constructor implemented as the state providers will be calling the default constructor to set the default state of the grain. If there is no default constructor, we will see a runtime exception.

The `ClearStateAsync` method should implement the logic to clean up the state by deleting the file from the persistent storage. We'll leave this as an exercise for you. You can refer to the sample code of this book on GitHub.

8. To follow the ASP .NET Core pattern for dependency injection, we need the extension methods to be added, which will be used to register all the required services. Covering this is beyond the scope of this book. Please refer to the sample code of this book on GitHub for reference.

With this, we have implemented a custom storage provider for Azure File storage. In the next section, use this provider to persist the grain state to Azure Files.

Grains with multiple states

An Orleans grain can have more than one state and they can be from different state providers. The `HotelGrain` class we implemented has the `checkedInGuests` state and it is persisted to Azure Cosmos DB. The main business of Distel is providing hospitality at all the locations of the seven wonders of the world. Consider that we may have a partnership with others to provide services such as transport for our guests. To manage this, we need to have the `HotelGrain` state manage the partner relationships. Though this should be part of the `HotelGrain` state, we may not want to club this with the `checkedInGuests` state as they both represent two different entities. Since the two states are separate concerns from each other, we will track them separately, with two different state providers. In this section, we will learn about having multiple states for the grain with different providers.

Follow the steps outlined here to add a new state:

1. Let's start by creating the model class with the name `Partner`, which defines the storage contract to persist the partner information. Add this class to the `Distel.Grains.Interfaces` project under the `Models` folder:

```
public class Partner
{
    public string Id { get; set; }
    public string Name { get; set; }
    public string PartnerType { get; set; }
}
```

2. Update the silo builder to use the custom `AzureFileGrainStorage` storage provider as shown in the following code snippet:

```
public static IHostBuilder CreateHostBuilder(string[]
    args) =>
    Host.CreateDefaultBuilder(args)
    .UseOrleans(siloBuilder =>
    {
        siloBuilder
        .AddCosmosDBGrainStorageAsDefault(opt =>
        {
            opt.AccountEndpoint =
                "https://localhost:8081";
            opt.AccountKey = "CosmosDB storage account
                Key";
            opt.DB = "distelstore";
        })
        .AddAzureFileGrainStorage("FileShare", opt => {
            opt.Share = "distelstore";
            opt.ConnectionString = "<<Azure Storage
                Connection String >>";
        })
        .Configure<ClusterOptions>(opts =>{})
        .Configure<EndpointOptions>(opts =>{});
    });
```

In the preceding code, we are registering the `AzureFileGrainStorage` class as a storage provider by calling the `AddAzureFileGrainStorage` extension method. We are naming the file share `distelstore`. The registration takes the connection string of the Azure storage. Recall that we registered the `CosmosDB` provider as the default storage provider. We registered the file share provider with the name `FileShare`. To use the file share provider, we should refer to the provider's registered name, `FileShare`.

> **Note**
>
> You can refer to the `AzureFileSiloBuilderExtensions` class for the implementation of the `AddAzureFileGrainStorage` extension method from the sample code of this book.

3. Add the `AssociatePartner` method to the interface definition of `IHotelGrain`:

```
public interface IHotelGrain : IGrainWithStringKey
{
    Task<string> WelcomeGreetingAsync(string
      guestName);
    Task AssociatePartner(Partner partner);
}
```

The preceding snippet defines the `AssociatePartner` method, which takes the `Partner` model that we created in the first step as an argument.

4. Let's now update the `HotelGrain` class to load the associated partners' state at the time of activation:

```
public class HotelGrain : Grain, IHotelGrain
{
    private IPersistentState<List<UserCheckIn>>
      checkedInGuests;
    private IPersistentState<List<Partner>> partners;
    public HotelGrain(ILogger<HotelGrain> logger,
        [PersistentState("checkedInGuests")]
    IPersistentState<List<UserCheckIn>>
      checkedInGuests,
        [PersistentState("partners", "FileShare")]
    IPersistentState<List<Partner>> partners)
```

```
        {
                this.checkedInGuests = checkedInGuests;
                this.partners = partners;
        }
    }
```

In the preceding code snippet, we added the `partners` constructor argument to hold the associated partners' state. The `PersistentState` attribute of this argument has two parameters. The first parameter denotes the name of the state, which is `partners`. The second argument specifies the named storage provider, `FileShare`, which is registered in the silo builder. The default storage provider will be picked if we do not specify the provider's name as in the case of `checkedInGuests`. The Orleans runtime will load the associated partners from the storage provider and populate the `partners` state object at the time of grain activation.

5. Implement the `AssociatePartner` method as shown in the following code snippet:

```
    public async Task AssociatePartner(Partner partner)
    {
      if (!this.partners.State.Any(e => e.Id ==
        partner.Id))
          {
                this.partners.State.Add(partner);
                await this.partners.WriteStateAsync();
          }
    }
```

In the preceding code snippet, when a request comes to associate a new partner, if there is no existing partner with same ID, we are updating the in-memory state to have the new partner and persisting it to the storage provider by calling the `WriteStateAsync` method. This method will persist the state in the Azure file share.

6. Add an action to `DistelController` to onboard a new partner as shown in the following code:

```
    [HttpPost("partner/{hotel}/onboard")]
    public async Task<IActionResult>
      OnboardPartner([FromRoute] string hotel, [FromBody]
```

```
        Partner partner)
    {
        var hotelGrain =
            this.clusterClient.GetGrain<IHotelGrain>(hotel);
        await hotelGrain.AssociatePartner(partner);
        return Ok();
    }
```

Here we are calling the `AssociatePartner` method on `HotelGrain` activation by passing in the partner details as an argument.

7. Now, if we run this solution and call the onboard API from Swagger UI, we will see the partners' state is persisted to the Azure file storage as shown in the following screenshot.

Figure 5.3 – Persisted grain state in Azure Files

As we defined in the Azure file storage providers, a file share will be created with the name `distelstore` specified in the silo configuration. A directory will be created with the name of the grain type. This directory will hold all the state objects for the Hotel grain type. The state will be serialized and stored in the file with its name as the grain identity.

In this section, we leveraged the customer persistence provider for Azure Files to persist the associated partners' state of the `HotelGrain`. While doing so, we learned how to use multiple storage providers to store the different states of the same grain.

You may be wondering if the state persistence should always use the state providers that we discussed in this chapter? The answer is *no*. The persistence provider system in Orleans is entirely optional. Developers may choose not to use Orleans persistence providers and go with directly accessing the database.

One of the questions that comes while designing the storage model is how much state can a grain have? The grain state will have the data required to perform operations or for future references. If the state contains hundreds or thousands of entities, the state grows unbounded, we need to think about splitting the state or directly accessing the database.

In the next section, we will learn about directly accessing the database from grains for state persistence.

Grains directly interacting with storage

The storage provider plugin model follows the pattern of loading the state at the time of activation. `IPersistentState` will be injected into the grain through the constructor. The state will be loaded before the `OnActivateAsync` method. The state will be persisted back to the storage provider by calling `WriteStateAsync` and can be refreshed at any time by calling the `ReadStateAsync` method. But this model may not work for all scenarios. In this section, we will see the scenarios that require the grain to access the database directly and ways to achieve that.

Consider the scenario where the write operation demands to call a stored procedure in the Cosmos DB provider, or we need to run a custom query to load the state on an ADO.NET provider. Both these functionalities may not be supported by the existing storage provider. In addition to this, having all that data as part of the grain state and loading it at the time of activation may be overkill and add latency.

The scenarios described above demand the need to access the database directly from the grain. There should be a provision to read and write the data from grains as we do in traditional N-tier applications with the data tier taking care of read-write operations.

We are co-hosting the Distel Orleans application along with the ASP.NET Core API. The services registered with `IServiceCollection` are available to the Orleans application as it shares the host builder of the ASP.NET Core application. We register a service as shown in the following code snippet by specifying the lifetime of the service:

```
public void ConfigureServices(IServiceCollection services)
{
    services.AddSingleton<IUserRepository, UserRepository>();
}
```

In the preceding code snippet, the `IUserRepository` service is registered with a singleton lifetime scope. The registered `IUserRepository` service can be injected into the grain via constructor injection as shown in the following code:

```
public class UserGrain : Grain, IUserGrain
{
    private readonly IUserRepository userRepository;

    public UserGrain(IUserRepository userRepository)
    {
        this.logger = logger;
        this.userRepository = userRepository;
    }

    public async Task<TravelHistory> GetTravelHistory()
    {
        var userId = this.GetPrimaryKeyString();
        var history = await
           userRepository.GetTravelHistoryAsync(userId);
        return history;
    }
}
```

In the above `UserGrain` implementation, `IUserRepository` is injected via the constructor. The injected `IUserRepository` service is used to fetch the travel history when the call comes to the `GetTravelHistory` grain method by calling `GetTravelHistoryAsync` on the injected service. The implementation of `IUserRepository` might be pulling the data from a SQL Server or any other service that is abstracted from the grain implementation.

This pattern can be used to execute any read or write operations from the grain method. There are two aspects that we should consider while using this pattern:

- The first aspect is the lifetime of the injected service. The lifetime of the grain is determined by the Orleans runtime. The services injected into the IServiceCollection container follow the lifetime specified at the time of service registration: *transient*, *scoped*, or *singleton*. If a transient service is injected via a constructor to the grain, then we might get some weird results as the service injected will be reused for the subsequent calls to the same grain activation. In Orleans, each grain activation gets a DI scope, so a "scoped" service will be created once per activation. That is, if the developer wants a per-grain service, then they can use scoped services. We will not have any problems consuming singleton services in grains. Also, services should not be passed in method calls unless they are serializable (which they mostly would not be).

- The second aspect that we should be considerate of is data consistency. Since data operations are not in the scope of the grain, a developer should handle the issues that may arise with concurrent operations and the updates that may happen to the data directly on the database by some other means.

In this section, we have seen how grain implementation can access the services injected into the **IoC** (**Inversion of Control**) container. Using this, we can interact with any service or datastore directly from the grain implementation.

Summary

In this chapter, we have learned about grain state and different providers to persist grain state. We used the Cosmos DB provider from the **OrleansContrib** community to persist our HotelGrain state. We implemented a custom storage provider for Azure file storage. We also learned how to use more than one provider in the same grain. Now you have the skills required to manage the state of distributed applications built with Microsoft Orleans. You also know how to create a custom state provider if there is a need.

Till now, all the interactions that we have built have been client-driven, which means we do the operation when a message is received from the client. There are situations where we need an operation to execute at regular intervals. In the next chapter, we will learn about reminders and timers in Orleans.

Further reading

- To learn more about Cosmos DB, refer to `https://gearup.microsoft.com/product/azure-cosmos-db`.

- To learn more about the Azure Files offering, refer to `https://azure.microsoft.com/en-in/services/storage/files/`.

- To learn more about `ShareClient` to manage Azure Files, refer to `https://docs.microsoft.com/en-us/dotnet/api/azure.storage.files.shares.shareclient?view=azure-dotnet`.

Questions

1. A grain can have only one state object.

   A. True

   B. False

   Answer – B

2. What is the attribute to be used on the grain constructor state parameter?

   A. `StorageState`

   B. `StatePesistence`

   C. `PersistentState`

   D. `State`

   Answer – C

3. Which interface defines the contract of an Orleans storage provider?

   A. `IStorage`

   B. `IGrainStorage`

   C. `IPersistentStorage`

   D. `IState`

   Answer – B

4. The Orleans runtime will call the `ReadStateAsync` method of the storage provider to load the state upon grain activation.

   A. True

   B. False

   Answer – A

5. Which method handles the cleanup of the grain state?

   A. `ClearStateAsync`

   B. `CleanStateAsync`

   C. `DeleteStateAsync`

   D. None

   Answer – A

6
Scheduling and Notifying in Orleans

It is a very common requirement in any application or product to execute a task/job at a specified date-time or a specified interval on a recurring schedule. A very common example of scheduling in our day-to-day life is alarms and reminders that we get from devices and apps such as phones, Outlook, and so on. The Orleans runtime provides us with two features, called **timers** and **reminders**, that enable the developer to achieve scheduling in grains.

It is also a very common requirement to send asynchronous notifications to clients. The Orleans runtime provides a feature called **observers** through which we can send notifications in grains.

This chapter will cover the following:

- Understanding and implementing timers
- Understanding and implementing reminders
- Understanding and implementing notifications

Technical requirements

To follow along, you will need to have the following:

- Visual Studio 2022 Community Edition

- A basic understanding of Azure and .NET

The code used in this chapter can be found at `https://github.com/ PacktPublishing/Distributed-.NET-with-Microsoft-Orleans/tree/ main/Chapter06`.

Understanding and implementing timers

The `HotelGrain` class that we have created in the book so far has been getting messages from external clients for welcome greetings or check-in as shown in the following diagram. Ultimately, all the `HotelGrain` classes have responded to an outside request.

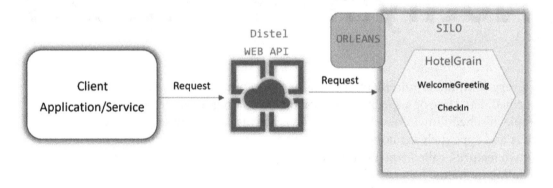

Figure 6.1 – HotelGrain responding to messages from external clients

But what if the grain needs to do some internal processing on a periodic basis. Timers provided by Orleans help in scheduling and creating periodic grain behavior without spanning multiple instances of the grain. Timers are asynchronous, single-threaded, and re-entrant so that a function can fire while the grain is awaiting. A grain can register multiple timers and can also cancel them when it is done with them. Timers are valid for the duration of the grain's activation, so they continue to fire until the grain is deactivated.

`Grain.RegisterTimer` shown here is the method to register a function to be invoked by the timer:

```
public IDisposable RegisterTimer(
        Func<object, Task> asyncCallback,
        // function invoked when the timer ticks
        object state,
        // object to pass to asyncCallback
        TimeSpan dueTime,
        // time to wait before the first timer tick
        TimeSpan period)
        // the period of the timer
```

This returns an `IDisposable` reference, which we can use to cancel the timer by disposing of it. In the next section, let's implement a timer in the `HotelGrain` class.

Implementing a timer in a HotelGrain

It is time for us to create our first timer. In this section, we are going to periodically display the number of available rooms. In a real-world scenario, you can assume this is a display of the available rooms on the hotel's rooftop digital banner. Whenever there is a new check-in, the available rooms will decrease by 1 and whenever there is a check-out, the available rooms will increase by 1.

Step 1: Create a callback function to be registered with a timer:

1. Add the following function to the `HotelGrain` class in the `Distel.Grains` project:

```
Task Callback(object callbackstate)
{

        Console.WriteLine("Total available
        rooms {0}", totalRooms -
        this.checkedInGuests.State.Count);

        return Task.CompletedTask;
}
```

2. For now, we can hardcode total number of rooms by adding it as a member variable with an initial value of 100 in the HotelGrain class:

```
private readonly int totalRooms = 100;
```

Step 2: Register the timer to send periodic callbacks to the function:

1. Add the following method to the HotelGrain class in the Distel.Grains project:

```
public override Task OnActivateAsync()
    {
            this.RegisterTimer(this.Callback,
            null, TimeSpan.FromSeconds(5),
            TimeSpan.FromSeconds(5));
            return base.OnActivateAsync();
    }
```

On the activation of the grain, the OnActivateAsync method will be invoked where the Callback function created in *Step 1* is now registered with the timer.

2. We don't have an object to pass to callback, hence we can set it to null, setting 5 seconds as the time to wait before the first timer tick and also setting 5 seconds for the recurring schedule and period of the timer tick.

Step 3: Testing the timer:

1. Check the checked-in guests count from the persistent storage Azure Cosmos DB implemented in the preceding chapter. In my case, it is two:

Figure 6.2 – Checked-in guests count from the persistent storage

2. Run the `Distel.WebHost` solution and activate `HotelGrain` by sending a welcome greeting from Swagger UI. After that, for every 5 seconds, you should see available rooms being printed in the console window as shown here.

```
info: Orleans.Runtime.SiloLifecycleSubject[100452]
      ClusterHealthMonitor started in stage Active (20000) in 0.2625 Milliseconds
info: Orleans.Runtime.SiloLifecycleSubject[100452]
      Orleans.Runtime.MembershipService.LocalSiloHealthMonitor started in stage Active (20000) in 0.1736 Milliseconds
info: Orleans.Runtime.SiloLifecycleSubject[100452]
      MembershipTableCleanupAgent started in stage Active (20000) in 0.2023 Milliseconds
info: Orleans.Runtime.SiloLifecycleSubject[100452]
      GatewayConnectionListener started in stage Active (20000) in 4.4665 Milliseconds
info: Orleans.Runtime.SiloLifecycleSubject[100452]
      Orleans.Runtime.Silo started in stage Active (20000) in 7.7142 Milliseconds
info: Orleans.Runtime.SiloLifecycleSubject[100452]
      Starting lifecycle stage Active (20000) took 22.5715 Milliseconds
info: Orleans.Hosting.SiloHostedService[0]
      Orleans Silo started.
info: Microsoft.Hosting.Lifetime[0]
      Application started. Press Ctrl+C to shut down.
info: Microsoft.Hosting.Lifetime[0]
      Hosting environment: Development
 WelcomeGreetingAsync message received: greeting = 'bhupesh'
Total available rooms 98
Total available rooms 98
Total available rooms 98
Total available rooms 98
Total available rooms 98
Total available rooms 98
Total available rooms 98
Total available rooms 98
Total available rooms 98
Total available rooms 98
```

Figure 6.3 – Available rooms printed by the timer callback

3. From Swagger UI, perform a new check-in as shown in the following screenshot to test if the available rooms count is decreasing:

Figure 6.4 – Perform a new check-in from Swagger UI

4. Check the updated available count being displayed now:

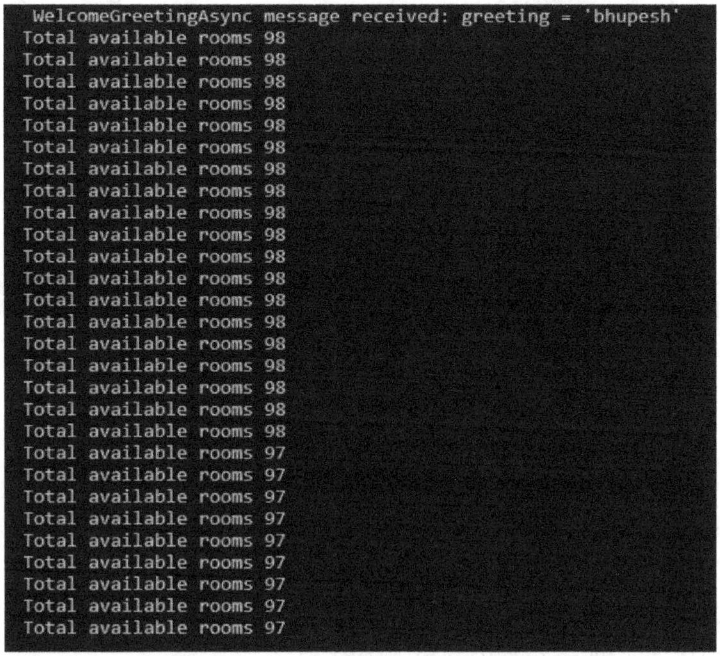

```
WelcomeGreetingAsync message received: greeting = 'bhupesh'
Total available rooms 98
Total available rooms 98
Total available rooms 98
Total available rooms 98
Total available rooms 98
Total available rooms 98
Total available rooms 98
Total available rooms 98
Total available rooms 98
Total available rooms 98
Total available rooms 98
Total available rooms 98
Total available rooms 98
Total available rooms 98
Total available rooms 98
Total available rooms 98
Total available rooms 98
Total available rooms 98
Total available rooms 97
Total available rooms 97
Total available rooms 97
Total available rooms 97
Total available rooms 97
Total available rooms 97
Total available rooms 97
Total available rooms 97
```

Figure 6.5 – Updated available rooms printed by the timer callback

As mentioned before, timers are valid for the duration of the grain's activation, so they continue to fire until the grain is deactivated. What if there is a need to fire periodic tasks beyond the deactivation of the grain? That's where reminders come into the picture.

Understanding and implementing reminders

Reminders are very similar to timers but with a few key differences:

- Reminders last beyond the life of the grain and are persistent.

- If the grain is deactivated before the reminder fires, the grain will be reactivated for the reminder.

- Even when there are cluster restarts, reminders are persistent and continue to trigger unless you cancel them explicitly.

- Reminders are single-threaded with asynchronous execution and by default will not process any incoming requests while the thread is awaiting for response from another call in progress unless you make it re-entrant.

- Reminders should not be used for high-frequency tasks as the period is measured in minutes, hours, or days whereas periods for timers are allowed in seconds. The minimum interval between reminders is 1 minute.

Configuration

Reminders need storage to function persistently. We must configure the storage for a reminder in silo builder via the `UseXReminderService` extension methods. Here, X is the provider's name, which can be `AzureTable` or `SQL`. Let's see how to configure the providers:

- **Azure Table provider**: You can configure `AzureTable` as storage for the reminder using the following code snippet:

```
// TODO replace with your connection string
const string connectionString =
  "YOUR_CONNECTION_STRING_HERE";
var silo = new SiloHostBuilder()
    [...]
    .UseAzureTableReminderService(options =>
    options.ConnectionString = connectionString)
    [...]
```

- **SQL provider**: You can configure SQL as storage for the reminder using the following code snippet:

```
// TODO replace with your connection string
const string connectionString =
  "YOUR_CONNECTION_STRING_HERE";
const string invariant = "YOUR_INVARIANT";
var silo = new SiloHostBuilder()
    [...]
    .UseAdoNetReminderService(options =>
    {
        options.ConnectionString =
            connectionString;
        options.Invariant = invariant;
```

```
    })
    [...]
```

- **In-memory provider for development box only**: You can configure `InMemory` as storage for the reminder using the following code snippet:

```
var silo = new SiloHostBuilder()
    [...]
    .UseInMemoryReminderService()
    [...]
```

Let's configure and implement a reminder in the next section.

Implementing reminders in a grain

It is time for us to create our first reminder. In this section, we are going to send a reminder at periodic intervals to clean the hotel. In a real-world scenario, you can assume this is a reminder sent to a person on the current shift to clean the hotel.

Let's get started:

1. *Step 1: Configuring storage for reminders*:

   Update the silo builder in `Program.cs` under the `Distel.WebHost` project to use the in-memory reminder service for this reminder demo:

```
<<Code removed for brevity>>
builder.Host.UseOrleans(siloBuilder =>
{
    siloBuilder
    .UseLocalhostClustering()
    .UseInMemoryReminderService()
    // Register Cosmos DB provider as the default
        stirage provider
    .AddCosmosDBGrainStorageAsDefault(opt =>
    {
        opt.AccountEndpoint = "<<Update account
            endpoint>>";
        opt.AccountKey = "<<Update cosmos DB
            Account Key>>";
        opt.DB = "distelstore";
        opt.CanCreateResources = true;
```

```
        })
<<Code removed for brevity>>
```

2. *Step 2: Implement a reminder interface in the grain:*

    All grains using the reminder functionality should implement IRemindable.
    ReceiveReminder. Hence, the HotelGrain class should implement the
    IRemindable.ReceiveReminder method. The reminderName string is a
    string that uniquely identifies the reminder within the scope of the contextual grain.
    DueTime specifies a duration of time to wait before issuing the first reminder tick.
    The period specifies the period of the timer between each reminder tick:

```
public class HotelGrain : Grain, IHotelGrain,
  IRemindable
{
<<Code removed for brevity>>

Task IRemindable.ReceiveReminder(string
  reminderName, TickStatus status)
{
//In real world reminder can be sent to on
  duty person in
//cleaning department
Console.WriteLine("Reminder to Clean the room");
  return Task.CompletedTask;
}

<<Code removed for brevity>>

}
```

3. *Step 3: Register the reminder to get a periodic reminder:*

    Add the following method to the HotelGrain class in the Distel.Grains
    project:

```
public override Task OnActivateAsync()
{
    disposableAvailableRoomTimer =
```

```
this.RegisterTimer(this.Callback, null,
    TimeSpan.FromSeconds(5),
    TimeSpan.FromSeconds(5));
this.RegisterOrUpdateReminder
    ("AvailableRoomCount",
        TimeSpan.FromSeconds(5),
        TimeSpan.FromMinutes(1));
    return base.OnActivateAsync();
}
```

On the activation of the grain, the `OnActivateAsync` method will be invoked and we will register the reminder there to invoke `Callback`. The `DueTime` method parameter specifies a quantity of time to wait before issuing the first timer tick and is specified as 5 seconds. The `Period` method parameter specifies the recurring period of the timer and is specified as 1 minute.

4. *Step 4: Testing the reminder*:

    Run the `Distel.WebHost` solution and activate the hotel grain by sending a welcome greeting from Swagger UI. After that, for every 5 seconds, you should see available rooms being printed in the console window by the timer and after every minute, you should see a reminder being printed in the console window.

```
ime.ReminderService.GrainBasedReminderTable/12345@fe52439f to 5127.0.0.1:11111:372093366*grn/Orleans.Runtime.ReminderSer
vice.GrainBasedReminderTable/12345@7dcfcdb6 after Duplicate activation. Attempt 0
info: Distel.Grains.HotelGrain[0]

 WelcomeGreetingAsync message received: greeting = 'bhupesh'
Total available rooms 97
Reminder to Clean the room
Total available rooms 97
Total available rooms 97
Total available rooms 97
Total available rooms 97
Total available rooms 97
Total available rooms 97
Total available rooms 97
Total available rooms 97
Total available rooms 97
Total available rooms 97
Reminder to Clean the room
Total available rooms 97
Total available rooms 97
Total available rooms 97
Total available rooms 97
Total available rooms 97
Total available rooms 97
Total available rooms 97
Total available rooms 97
■
```

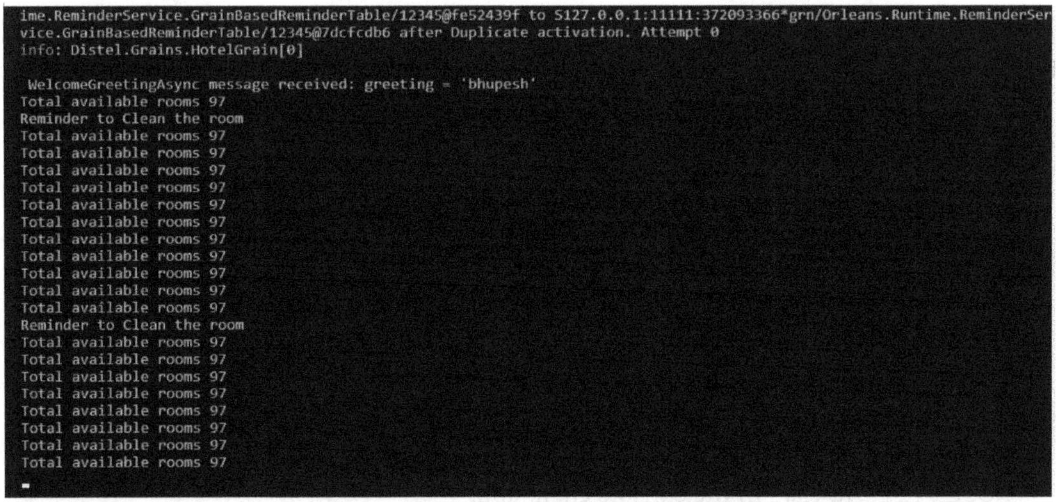

Figure 6.6 – Reminder sent to clean the room

We created a timer and reminder and tested them. In the next section, we will see how we can send notifications to clients asynchronously.

Understanding and implementing notifications

There are several scenarios in the real world where clients need to receive asynchronous notifications. For example, when a check-in happens, the available room count reduces, and we need to notify all clients who might have enabled booking for our hotel on their sites to show current availability. Orleans provides a feature called **client observers** that helps to notify clients asynchronously. Let's configure and implement notifications in the next sub-section.

Implementing notifications in a grain

The mechanisms that allow a grain to notify clients asynchronously are called observers. An observer is an interface that inherits from `IGrainObserver` and the client implementing `IObserver` is called client observers. The grain will provide an API to subscribe or unsubscribe observers and send notifications to subscribed observers. Let's implement observer notifications step by step.

Step 1: Configuring the client to receive notifications:

1. Define the interface for the client that will receive the message. Add the following interface to the `Distel.Grains.Interfaces` project:

    ```
    public interface IObserver : IGrainObserver
    {
            void ReceiveMessage(string message);
    }
    ```

    The important rule is that the interface should inherit from `IGrainObserver`. The name can be anything. In this case, we have kept it as `Iobserver`. The message in our example is `string`.

2. The client who wants to receive notifications by being an observer needs to implement a class that implements `IObserver`. Add the following code to the `HotelClient` class in the console client app:

    ```
    public class HotelClient : IObserver
    {
            public void ReceiveMessage(string
    ```

```
        message)
    {
        Console.WriteLine(message);
    }
}
```

Step 2: Configuring grains to enable clients to subscribe for and unsubscribe from notifications:

1. IHotelGrain should expose two methods, Subscribe and Unsubscribe:

    ```
    public interface IHotelGrain :
      IGrainWithStringKey
        {
            <<Code removed for brevity>>
            public Task Subscribe(Iobserver
                observer);
            public Task UnSubscribe(Iobserver
                observer);
        }
    ```

2. The ObserverManager helper class is provided in Orleans, which internally manages the subscribers for us. Create a new class, ObserverManager.cs, and add it to the Distel.Grains project. You can find the complete code here: https://github.com/PacktPublishing/Distributed-.NET-with-Microsoft-Orleans/tree/main/Chapter06/Distel/Distel.Grains/ObserverManager.cs. This helper class has methods to manage the Subscribe and Unsubscribe methods and Notify as shown here:

    - Subscribe: Ensures that the provided observer is subscribed, renewing its subscription:

    ```
    public class ObserverManager<TObserver> :
      ObserverManager<IAddressable, TObserver>
        {
            public void Subscribe(TAddress address,
                TObserver observer)
            {
                // Add or update the subscription.
    ```

```
var now = this.GetDateTime();
ObserverEntry entry;
if
  (this.observers.TryGetValue
  (address, out entry))
{
    entry.LastSeen = now;
    entry.Observer = observer;
    if
      (this.log.IsEnabled
      (LogLevel.Debug))
    {
      this.log.LogDebug
      (this.logPrefix + ":
        Updating entry
        for {0}/{1}. {2}
        total subscribers.",
        address, observer,
        this.observers.Count);
    }
}
else
{
    this.observers[address] = new
    ObserverEntry { LastSeen = now,
      Observer = observer };
    if
      (this.log.IsEnabled
      (LogLevel.Debug))
    {
      this.log.LogDebug
      (this.logPrefix + ": Adding
        entry for {0}/{1}. {2} total
        subscribers after add.",
        address, observer,
        this.observers.Count);
```

```
                    }
                }
            }
    }
```

- `Unsubscribe`: Ensures that the provided subscriber is unsubscribed:

```
public void Unsubscribe(TAddress subscriber)
        {
            this.log.LogDebug(this.logPrefix + ":
              Removed entry for {0}. {1} total
              subscribers after remove.",
              subscriber, this.observers.Count);
            this.observers.TryRemove(subscriber,
              out _);
        }
```

- `Notify`: Ensures that all observers are notified:

```
public void Notify(Action<TObserver>
  notification, Func<TObserver, bool> predicate =
    null)
        {
            var now = this.GetDateTime();
            var defunct =
              default(List<TAddress>);
            foreach (var observer in
              this.observers)
            {
                if (observer.Value.LastSeen +
                  this.ExpirationDuration < now)
                {
                    // Expired observers will be
                      removed.
                    defunct = defunct ?? new
                      List<TAddress>();
                    defunct.Add(observer.Key);
```

```
                    continue;
        }


        // Skip observers which don't
          match the provided predicate.
        if (predicate != null &&
          !predicate(observer.Value.
          Observer))
        {
            continue;
        }
        try
        {
         Notification
            (observer.Value.Observer);
        }
        catch (Exception)
        {
            // Failing observers are
              considered defunct and
              will be removed..
            defunct = defunct ?? new
              List<TAddress>();
            defunct.Add(observer.Key);
        }
    }


    // Remove defunct observers.
    if (defunct !=
      default(List<TAddress>))
    {
        foreach (var observer in defunct)
        {
         this.observers.TryRemove
            (observer, out _);
            if
```

```
                    (this.log.IsEnabled
                    (LogLevel.Debug))
            {
                    this.log.LogDebug
                    (this.logPrefix + ":
                    Removing defunct entry for
                    {0}. {1} total subscribers
                    after remove.", observer,
                    this.observers.Count);
            }
        }
    }
}
```

3. HotelGrain should implement two methods, Subscribe and Unsubscribe.
    Internally, it will use the Subscribe and Unsubscribe methods of the
    ObserverManager instance:

```
public class HotelGrain : Grain, IHotelGrain,
    IRemindable
    {
        <<Code removed for brevity>>
        private readonly
            ObserverManager<IObserver>
            observerManager;

        public HotelGrain(ILogger<HotelGrain>
            logger,[PersistentState
            ("checkedInGuests")]
        IPersistentState<List<UserCheckIn>>
            checkedInGuests)
        {
            <<Code removed for brevity>>
            observerManager = new
                ObserverManager<IObserver>
                (TimeSpan.FromMinutes(5), logger,
                "subs");
```

```
        }

        // Clients call this to subscribe.
        public Task Subscribe(IObserver observer)
        {
            observerManager.Subscribe(observer,
                observer);
            return Task.CompletedTask;
        }

        //Clients call this to unsubscribe
        public Task UnSubscribe(Iobserver
          observer)
        {
            observerManager.Unsubscribe(observer);
            return Task.CompletedTask;
        }
    <<Code removed for brevity>>
    }
```

Step 3: Sending notifications from grains:

Add the SendUpdateMessage method to the HotelGrain class in the
Distel.Grains project. We can leverage the notifying capability in the
ObserverManager utility class to notify all clients. For testing purposes, we can call the
SendUpdateMessage method from the Callback method invoked by the timer as
shown here:

```
public class HotelGrain : Grain, IHotelGrain,
  IRemindable
    {
<<Code removed for brevity>>

//Send message to subscribed clients
        public Task SendUpdateMessage(string message)
        {
            observerManager.Notify(s =>
                s.ReceiveMessage(message));
```

```
        return Task.CompletedTask;
    }


Task Callback(object callbackstate)
    {
        Console.WriteLine("Total available rooms
        {0}", totalRooms -
        this.checkedInGuests.State.Count);
        SendUpdateMessage("Total available rooms "
        + (totalRooms -
        this.checkedInGuests.State.Count)
        .ToString());
        return Task.CompletedTask;
    }
<<Code removed for brevity>>
```

Step 4: Client subscribing for notifications:

Create an observer reference in `ConsoleClientApp` using the `HotelClient` class, which implemented `IObserver`. To send notifications to all the subscribed clients, the `HotelGrain` class on the server should invoke the `SendUpdateMessage` method. In the client code, the `HotelClient` instance in the `hc` variable will receive the notification:

```
class Program
{
<<Code removed for brevity>>
private static async Task
  SendWelcomeGreeting(IClusterClient client, string
  guest)
        {
var hotel = client.GetGrain<IHotelGrain>("Taj");
        Console.WriteLine("Identity String : " +
            hotel.GetGrainIdentity().IdentityString);
var response = await
  hotel.WelcomeGreetingAsync(guest);
        Console.WriteLine($"\n\n{response}\n");
```

```
HotelClient hc = new HotelClient();
var obj = await
    client.CreateObjectReference
    <IObserver>(hc);
await hotel.Subscribe(obj);
    }
}
```

Step 5: Testing the notifications:

Run `Distel.WebHost` and the console client solution. We have added the code to subscribe for notifications in the `SendWelcomeGreeting` method. Once you run the console client solution, it will ask you to enter the guest name. Once you enter the guest name, it will invoke the `SendWelcomeGreeting` method where the code to subscribe for notifications also gets invoked. After that, you can see the client receiving notifications on the number of available rooms as shown here:

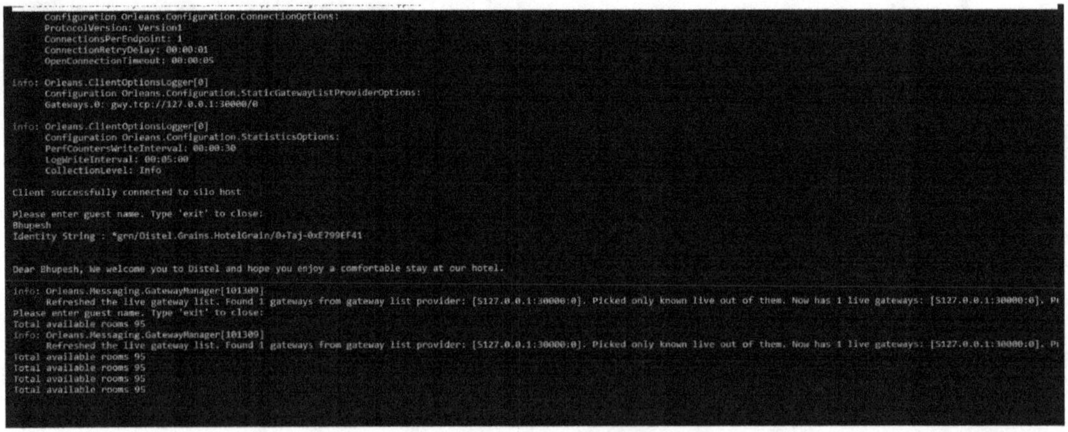

Figure 6.7 – Notification sent to the subscriber about available rooms

With this, we implemented an observer notification by configuring grains to enable clients to subscribe for and unsubscribe from notifications, sent notifications from grains, and looked at a client subscribing to receive notifications.

Summary

In this chapter, we discussed how to implement scheduling using timers leveraging the `Grain.RegisterTimer` method, which requires us to pass a callback function, a timespan to wait before the first timer tick, and finally a timespan to wait between each schedule of the timer. We also discussed how to implement scheduling using reminders leveraging the `Grain.RegisterOrUpdateReminder` method, which requires us to pass the reminder name, the timespan to wait before the first reminder tick, and the timespan to wait between each schedule of the reminder. We also discussed how to implement notifications using observers leveraging the `Grain.Subscribe`, `Grain.Unsubscribe`, and `Grain.Notify` methods. You will now be able to implement scheduling using timers and reminders and implement notifications using observers in Orleans.

In the next chapter, we will learn in detail about engineering fundamentals in Orleans.

Questions

1. How can we start a timer?

    A. By calling `Grain.Subscribe`

    B. By creating a new task

    C. By calling `Grain.RegisterTimer`

    D. None of the above

    Answer – C

2. How can we cancel a timer?

    A. By calling `Grain.CancelTimer`

    B. By disposing of it

    C. By calling `Grain.UnregisterTimer`

    D. None of the above

    Answer – B

3. How can we start a reminder?

    A. By calling `Grain.RegisterOrUpdateReminder`

    B. By calling `Grain.GetReminder`

    C. By calling `Grain.StartReminder`

    D. None of the above

    Answer – A

4. When can I use a timer?

    A. When there is no significant impact on functionality when the timer stops due to grain deactivation or the cluster restarts

    B. For performing frequent tasks in seconds or minutes

    C. Both A and B

    D. None of the above

    Answer – C

5. When can I use reminders?

    A. To perform infrequent tasks in minutes, hours, or days

    B. For tasks that should continue to trigger even when there is a partial or full cluster restart

    C. Both A and B

    D. None of the above

    Answer – C

7

Engineering Fundamentals in Orleans

As a developer, it is very important to understand your application inside out. Real-time dashboards provide insights into and metrics of what is happening inside your application and can be used by developers and admins to monitor the health of your application.

Once development is completed, you need to test your code thoroughly to identify issues early in the cycle and ensure quality gates are met before we ship the application to production. That's where quality unit testing plays a big part.

Once the application is shipped to production, the collection of data and logs such as user requests, warnings, and errors as part of telemetry is very important so that you can monitor and detect anomalies and raise alerts with support engineers on call to detect issues quickly and mitigate them before customers are impacted. That's where real-time telemetry comes into the picture.

In this chapter, we will learn about the following engineering fundamentals in Orleans:

- Setting up the Orleans dashboard
- Understanding unit testing in Orleans
- Real-time telemetry

Technical requirements

To follow along with the chapter, you should have the following:

- Visual Studio 2022 Community Edition
- An understanding of building .NET applications
- An Azure subscription

The code used in this chapter can be found at `https://github.com/PacktPublishing/Distributed-.NET-with-Microsoft-Orleans/tree/main/Chapter07`.

Setting up the Orleans dashboard

The Orleans dashboard helps developers to understand what is happening inside an Orleans application through simple metrics and insights. It is simple to set up – let's see how it can be configured and launched.

Step 1: Install the OrleansDashboard NuGet package and configure a silo:

1. Install the `OrleansDashboard` NuGet package in the `Distel.WebHost` project.

2. Add configuration as shown here in the `Distel.WebHost\Program.cs` file for `siloBuilder` to use the dashboard:

```
<<Code removed for brevity>>
builder.Host.UseOrleans(siloBuilder =>
{
        siloBuilder
        .ConfigureApplicationParts
            (parts => parts.AddFromApplicationBaseDirectory())
        .UseDashboard(options => { })
        .UseLocalhostClustering()
    .UseInMemoryReminderService()


    <<Code removed for brevity>>
```

> **Note**
>
> The dashboard uses `ConfigureApplicationParts` to register its services and grains, which disables the automatic discovery of grains in Orleans. To enable automatic discovery of the grains of the original project, change the configuration to use `.ConfigureApplicationParts(parts => parts.AddFromApplicationBaseDirectory())`.

Step 2: Start the silo and open the dashboard:

Run the `Distel.WebHost` project to start the silo, and open `http://localhost:8080` in your browser to see the dashboard as shown here:

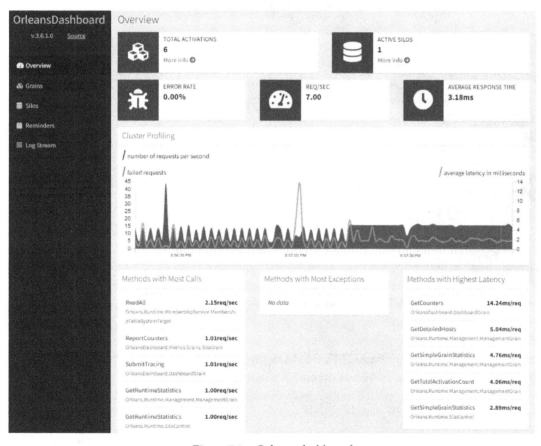

Figure 7.1 – Orleans dashboard

With this, the Orleans dashboard is available for developers and admins for monitoring. In the next two subsections, we will look at different features and APIs available in the dashboard respectively.

Dashboard features

Let's have a look at the different features provided by Orleans dashboard to understand what is happening inside Orleans application.

Grains – Activations by Type

In this section, you can see the total grain activations count and for each grain, you can see **Activations**, **Exception rate**, **Throughput**, and **Latency** details, as shown here:

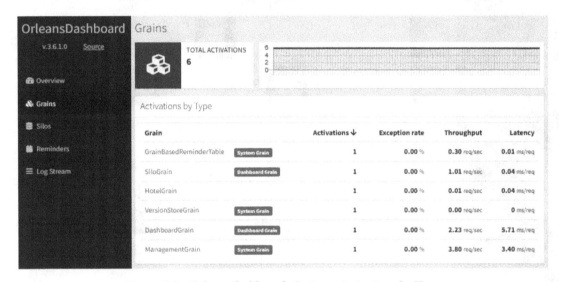

Figure 7.2 – Orleans dashboard: Grains – Activations by Type

Silos

In this section, you can see the **ACTIVE SILOS** total and **Silo Health**, such as the duration for which the silo has been up and active and the number of **activations**, as shown here:

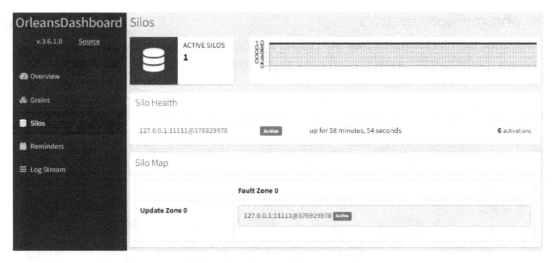

Figure 7.3 – Orleans dashboard: Silos

Reminders

In this section, you can see a **REMINDERS COUNT** total and can filter by **Grain Reference**, **Primary Key**, **Name**, **Start At**, and **Period**.

Figure 7.4 – Orleans dashboard: Reminders

CPU and memory metrics

In Windows, the CPU and memory metrics are enabled when you add the `Microsoft.Orleans.OrleansTelemetryConsumers.Counters` package and add `builder.UsePerfCounterEnvironmentStatistics()`. After some time, you will be able to see **CPU Usage**, **Memory Usage**, and **Grain Usage** percentages as shown in the following screenshot:

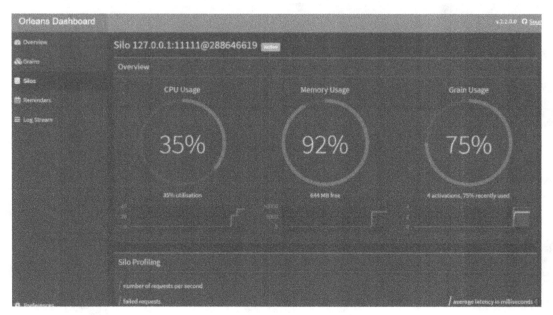

Figure 7.5 – CPU and memory usage

In the next subsection, we will look at the different APIs available for us to consume from the Orleans dashboard.

Dashboard APIs

Orleans dashboard exposes HTTP APIs that we can consume programmatically to get metrics and insights into what is happening inside your Orleans application. Let's have a look at the available APIs.

Dashboard Counters

`GET/DashboardCounters` returns a summary of cluster metrics. The summary contains total active hosts count currently and a history of total active hosts till now (`totalActiveHostsCountHistory`), total activations currently and a history of total activations till now (`totalActivationCountHistory`).

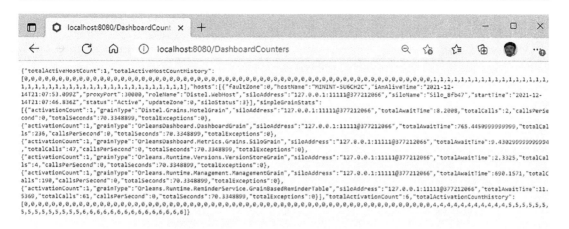

Figure 7.6 – Orleans dashboard Dashboard Counters API

Grain Stats

GET/GrainStats/{grainName} returns the profiling counters for the grain method
collected over the last 100 seconds for each of the grains, aggregated across all silos
as shown below. You will get grain name, method name, total counts to the grain and
method, total exceptions et cetera for the specific grain.

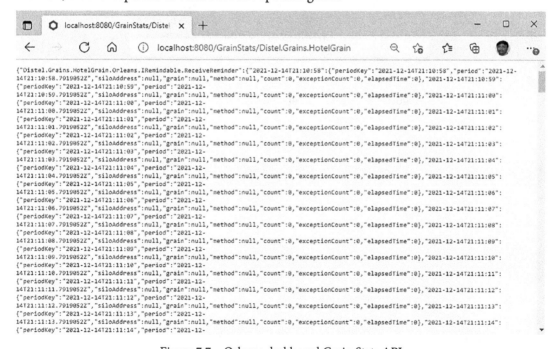

Figure 7.7 – Orleans dashboard Grain Stats API

Cluster Stats

GET/ClusterStats returns the aggregated profiling counters for the grain method collected over the last 100 seconds for the whole cluster as shown below. You will get grain name, method name, total counts to the grain and method, total exceptions et cetera for the whole cluster.

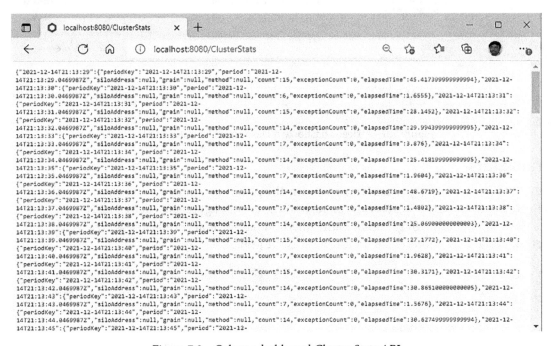

Figure 7.8 – Orleans dashboard Cluster Stats API

Silo Stats

`GET /SiloStats/{siloAddress}` returns the aggregated grain method profiling counters collected over the last 100 seconds for that silo.

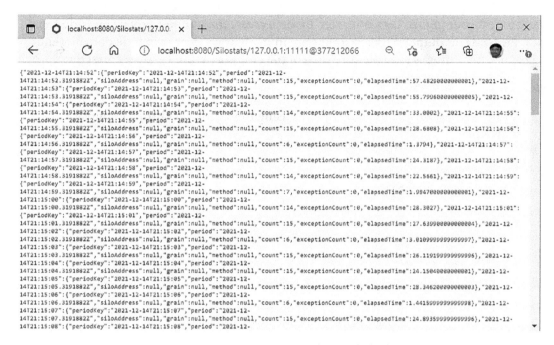

Figure 7.9 – Orleans dashboard Silo Stats API

With this, we are done with the Orleans dashboard features and APIs. By now, you should have a good understanding of setting up a dashboard to monitor and leverage the APIs exposed. Let's now spend some time looking at common enterprise architectures. In the next section, we will see how to do unit testing in Orleans.

Understanding unit testing in Orleans

In this section, we will see how to do unit testing on Orleans grains and ensure they are working fine. Orleans has a NuGet package called `Microsoft.Orleans.TestingHost`, which can be used to create an in-memory cluster with two silos by default to test your grains. Let's now, step by step, create the unit test project and add tests to test the `Grains` output. Orleans grains are in many ways their own little domains and the goal of unit testing is to make sure each of the grains as a unit behaves correctly.

Step 1: Create the unit test project:

1. Add a unit test project to your solution using the **xUnit Test Project** template as shown here:

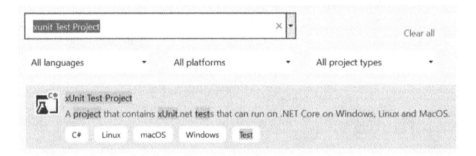

Figure 7.10 – Unit test project template

2. Name the unit test project `Distel.UnitTests`:

Figure 7.11 – Unit test project template

Step 2: Install the TestingHost NuGet package and configure the silo:

1. Install the `Microsoft.Orleans.TestingHost` NuGet package in the `Distel.UnitTests` project as shown here:

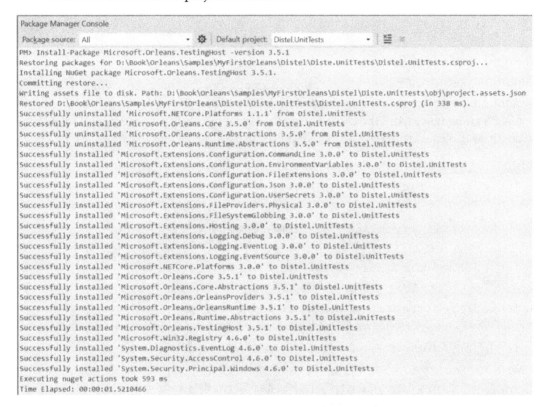

```
Package Manager Console

Package source:  All              ▾  ⚙  Default project:  Distel.UnitTests        ▾  ⬚

PM> Install-Package Microsoft.Orleans.TestingHost -version 3.5.1
Restoring packages for D:\Book\Orleans\Samples\MyFirstOrleans\Distel\Diste.UnitTests\Distel.UnitTests.csproj...
Installing NuGet package Microsoft.Orleans.TestingHost 3.5.1.
Committing restore...
Writing assets file to disk. Path: D:\Book\Orleans\Samples\MyFirstOrleans\Distel\Diste.UnitTests\obj\project.assets.json
Restored D:\Book\Orleans\Samples\MyFirstOrleans\Distel\Diste.UnitTests\Distel.UnitTests.csproj (in 338 ms).
Successfully uninstalled 'Microsoft.NETCore.Platforms 1.1.1' from Distel.UnitTests
Successfully uninstalled 'Microsoft.Orleans.Core 3.5.0' from Distel.UnitTests
Successfully uninstalled 'Microsoft.Orleans.Core.Abstractions 3.5.0' from Distel.UnitTests
Successfully uninstalled 'Microsoft.Orleans.Runtime.Abstractions 3.5.0' from Distel.UnitTests
Successfully installed 'Microsoft.Extensions.Configuration.CommandLine 3.0.0' to Distel.UnitTests
Successfully installed 'Microsoft.Extensions.Configuration.EnvironmentVariables 3.0.0' to Distel.UnitTests
Successfully installed 'Microsoft.Extensions.Configuration.FileExtensions 3.0.0' to Distel.UnitTests
Successfully installed 'Microsoft.Extensions.Configuration.Json 3.0.0' to Distel.UnitTests
Successfully installed 'Microsoft.Extensions.Configuration.UserSecrets 3.0.0' to Distel.UnitTests
Successfully installed 'Microsoft.Extensions.FileProviders.Physical 3.0.0' to Distel.UnitTests
Successfully installed 'Microsoft.Extensions.FileSystemGlobbing 3.0.0' to Distel.UnitTests
Successfully installed 'Microsoft.Extensions.Hosting 3.0.0' to Distel.UnitTests
Successfully installed 'Microsoft.Extensions.Logging.Debug 3.0.0' to Distel.UnitTests
Successfully installed 'Microsoft.Extensions.Logging.EventLog 3.0.0' to Distel.UnitTests
Successfully installed 'Microsoft.Extensions.Logging.EventSource 3.0.0' to Distel.UnitTests
Successfully installed 'Microsoft.NETCore.Platforms 3.0.0' to Distel.UnitTests
Successfully installed 'Microsoft.Orleans.Core 3.5.1' to Distel.UnitTests
Successfully installed 'Microsoft.Orleans.Core.Abstractions 3.5.1' to Distel.UnitTests
Successfully installed 'Microsoft.Orleans.OrleansProviders 3.5.1' to Distel.UnitTests
Successfully installed 'Microsoft.Orleans.OrleansRuntime 3.5.1' to Distel.UnitTests
Successfully installed 'Microsoft.Orleans.Runtime.Abstractions 3.5.1' to Distel.UnitTests
Successfully installed 'Microsoft.Orleans.TestingHost 3.5.1' to Distel.UnitTests
Successfully installed 'Microsoft.Win32.Registry 4.6.0' to Distel.UnitTests
Successfully installed 'System.Diagnostics.EventLog 4.6.0' to Distel.UnitTests
Successfully installed 'System.Security.AccessControl 4.6.0' to Distel.UnitTests
Successfully installed 'System.Security.Principal.Windows 4.6.0' to Distel.UnitTests
Executing nuget actions took 593 ms
Time Elapsed: 00:00:01.5210466
```

Figure 7.12 – Test Explorer

2. If you can recall from *Chapter 5*, *Persistence in Grains*, we made `HotelGrain` use the state from the persistent storage provider by adding Cosmos DB storage to the silo builder and in *Chapter 6*, *Scheduling and Notifying in Orleans*, we used an in-memory reminder service for reminder storage. For the unit test, we can add memory storage by leveraging the `AddMemoryGrainStorageAsDefault` and `UseInMemoryReminderService` methods available in `ISiloBuilder` as shown here:

```
public class TestSiloConfigurations :
    ISiloConfigurator
{
    public void Configure(ISiloBuilder siloBuilder)
```

```
        {
            siloBuilder.AddMemoryGrainStorageAsDefault();
            siloBuilder.UseInMemoryReminderService();
        }
    }
```

Step 3: Add the code to test:

Create a new instance of `TestClusterBuilder` and call the Build API, which will
return the cluster instance as shown in the following code snippet. We can get the
`HotelGrain` instance from the cluster's `GrainFactory.GetGrain` API and then call
the functions in `HotelGrain` as shown here:

```
public class HotelGrainTest
{
    [Fact]
    public async Task WelcomeGreeting()
    {
        var builder = new TestClusterBuilder();
        builder.AddSiloBuilderConfigurator
            <TestSiloConfigurations>();
        var cluster = builder.Build();
        cluster.Deploy();

        var hotelGrain = cluster.GrainFactory
            .GetGrain<IHotelGrain>("Taj");
        var greeting = await hotelGrain.
            WelcomeGreetingAsync("Bhupesh");

        cluster.StopAllSilos();
        Assert.Equal("Dear Bhupesh, We welcome you to
            Distel and hope you enjoy a comfortable
            stay at our hotel. ", greeting);
    }
}
```

Step 4: Run the test:

Build the unit test project and you can choose **Test** on the Visual Studio menu, and then choose **Test Explorer**. In **Test Explorer**, you can group tests into categories and create, save, and run playlists of tests. You can also use **Test Explorer** to debug unit tests. Run the **WelcomeGreeting** test as shown here.

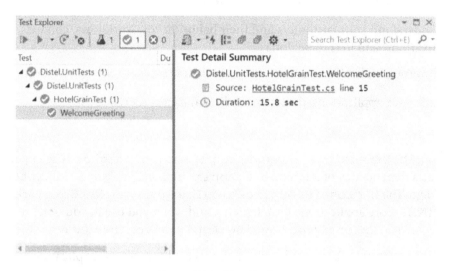

Figure 7.13 – Test Explorer

> **Note**
>
> Bringing up a test cluster *for each test* is expensive and *potentially* prone to runtime errors due to the nature of how the test cluster(s) is brought up. Here is an example from `OrleansContrib` to reuse a `TestFixture` base class to accomplish the reuse of an in-memory `TestCluster`: https://github.com/OrleansContrib/Orleans.SyncWork/tree/main/test/Orleans.SyncWork.Tests/TestClusters.

In this section, we have learned how to do unit testing on grains by creating a unit test project, adding the `Microsoft.Orleans.TestingHost` NuGet package, configuring the silo, and using `TestCluster`. In the next section, we will see how to add real-time telemetry to our applications.

Adding real-time telemetry

Application Insights is one of the best telemetry offerings, provided by Microsoft Azure for developers and DevOps professionals as an extensible **Application Performance Management (APM)** service to do the following:

- Derive data-driven insights with the following aims:

  - To understand your customers and their habits, needs, workflows, contexts, motivations, and jobs to be done

  - To improve application usability and in turn, increase customer satisfaction

- Monitor your applications, detect anomalies, and raise alerts.

The `Microsoft.Extensions.Logging.ApplicationInsights` NuGet package is included as a dependency of `Microsoft.ApplicationInsights.AspNetCore` NuGet package. The `Microsoft.ApplicationInsights.AspNetCore` package is used in ASP.NET Core applications for telemetry, and when you use it, you don't need to install `Microsoft.Extensions.Logging.ApplicationInsights`.

As shown in the following figure, you can install this package in your application to enable and write telemetry:

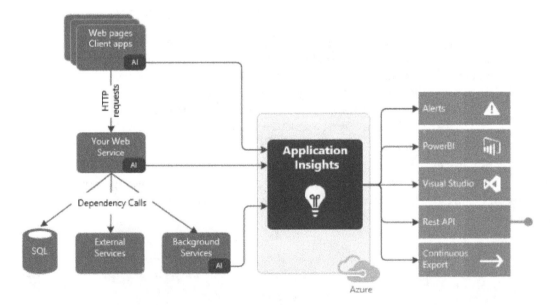

Figure 7.14 – Application Insights instrumentation for telemetry

> **Note**
>
> There is no impact on your application's performance due to instrumentation. Calls to Application Insights from your application are non-blocking, sent in a separate thread in batches.

Let's see how we can enable logging in our application using Application Insights.

Enabling application logging in Application Insights

The steps to enable application logging when using Application Insights are as follows:

1. Install the `Microsoft.ApplicationInsights.AspNetCore` package from **Tools | NuGet Package Manager | Package Manager Console** as shown, using the `Install-Package <Package name> -version <Version number>` command:

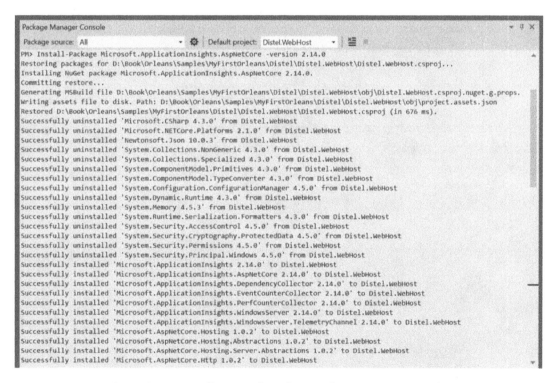

Figure 7.15 – Installing a package from Package Manager Console

2. After installing the package, you need to add the `Telemetry` section and update the instrumentation key (GUID) of your Azure Application Insights resource in `appsettings.json`, so that all telemetry data is written to your Azure Application Insights resource. If you don't have an Azure Application Insights resource, go ahead and create one and then add it to `appsettings.json`:

```json
{
    "Logging": {
      "LogLevel": {
        "Default": "Information",
        "Microsoft": "Warning",
        "Microsoft.Hosting.Lifetime": "Information"
      }
    },
    "Telemetry": {
      "InstrumentationKey": "Your AppInsights
      Instrumentation Key "
    }
}
```

3. In your `Distel.WebHost` project, `Program.cs` class, add the following code:

```
services.AddApplicationInsightsTelemetry
    ("Your AppInsights Instrumentation Key ");
```

4. Now you can build and run the `Distel.WebHost` application. You can see the telemetry data by navigating to your **Application Insights | Overview** where you can see **Failed requests, Server response time**, and **Server requests** as shown in the following screenshot:

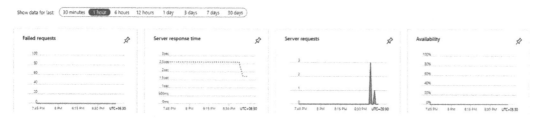

Figure 7.16 – Application Insights overview

5. You can navigate to **Application Insights | Live metrics** for real-time performance counters as shown in the following figure:

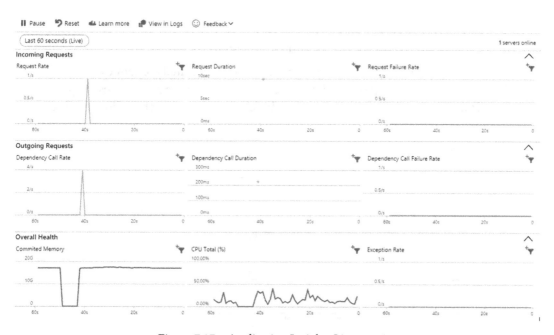

Figure 7.17 – Application Insights Live metrics

6. You can navigate to **Application Insights | Metrics** to get different metrics and charts, as shown in the following figure:

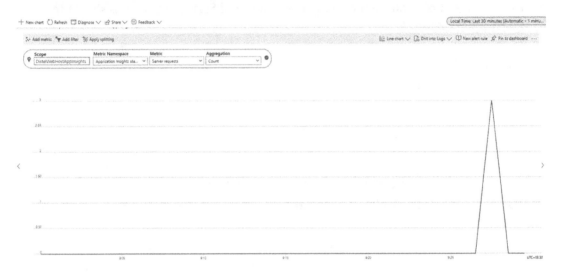

Figure 7.18 – Application Insights Metrics

7. You can navigate to **Application Insights | Failures** and analyze **Operations, COUNT (FAILED), Dependencies, Exceptions, Top 3 response codes, Top 3 exception types,** and **Top 3 failed dependencies,** as shown in the following figure:

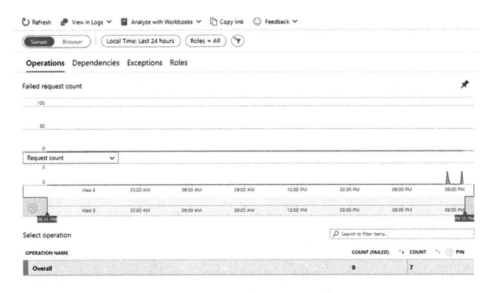

Figure 7.19 – Application Insights Failures

We've now seen Application Insights available out of the box. How do we add custom logs in the case of a warning or error? We can use the logger. (Object creation happens using **Dependency Injection (DI)**. DI is enabled in the second and third steps.) Let's add a custom log to the `WelcomeGuest` API using the following code:

```
[HttpGet("welcome/{hotel}/{guestname}")]
        public async Task<IActionResult>
          WelcomeGuest([FromRoute] string hotel,
          [FromRoute] string guestname)
        {
            var hotelGrain = this.clusterClient
              .GetGrain<IHotelGrain>(hotel);
            var greeting = await hotelGrain
              .WelcomeGreetingAsync(guestname);
            _logger.LogWarning("Warning: Welcome Guest
              successful for guest " + guestname +
              "for hotel " + hotel);
            _logger.LogError("Error: Welcome Guest
              successful for guest " + guestname +
              "for hotel " + hotel);
            return Ok(greeting);

        }
```

You can invoke the `WelcomeGuest` API from Swagger and then navigate to **Application Insights | Logs** and check traces where you can see both the warning and the error that we logged, as shown in the following figure:

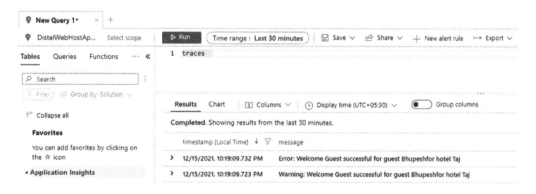

Figure 7.20 – Application Insights Logs

Application Insights is a very simple-to-use and powerful log provider. We have seen the rich telemetry it provides out of the box.

Summary

In this chapter, we introduced some of the engineering fundamentals of Orleans. We learned about the Orleans dashboard, its features, and the APIs available, which can be used to understand what is happening in your application. Using `Microsoft. Orleans.TestingHost`, we saw how our Distel application can be unit tested to identify issues upfront before shipping the code for E2E testing and production. We also learned how to enable real-time telemetry, using which you will be able to derive data-driven insights and conduct monitoring to detect anomalies and raise alerts.

In the next chapter, we'll learn about advanced concepts in Orleans such as streaming, heterogeneous silos, and grain interface versioning.

Questions

1. What are the features available in the Orleans dashboard?

    A. Grains – Activations by Type

    B. Silos

    C. Reminders

    D. All the above

    Answer – D

2. What are the APIs available in the Orleans dashboard?

    A. DashboardCounters

    B. GrainStats

    C. ClusterStats

    D. SiloStats

    E. All the above

    Answer – E

3. Which NuGet package can be used to create an in-memory cluster for unit testing in Orleans?

    A. `OrleansDashboard`

    B. `xUnit.Test`

    C. `Microsoft.Orleans.TestingHost`

    D. All the above

    Answer – C

4. Application Insights helps to:

    A. Derive data-driven insights.

    B. Understand your customers and their habits, needs, workflows, contexts, motivations, and jobs to be done.

    C. Improve application usability and in turn, increase customer satisfaction.

    D. Monitor your applications, detect anomalies, and raise alerts.

    E. All the above.

    Answer – E

Section 3 - Building Patterns in Orleans

In this section, we will learn about a few advanced concepts in Orleans, such as streaming and heterogeneous silos. This section will also cover some of the most popular patterns that are used while creating apps with Orleans.

In this section, we will cover the following topics:

8
Advanced Concepts in Orleans

In the last two parts of this book, you have learned about building a highly scalable distributed application using Orleans. We have also seen how to execute periodic tasks using Orleans timers and reminders, and you have learned how to write unit tests for an Orleans application. So, we have all the skills required to build a distributed application using Orleans. In this part of the book, we will delve into some of Orleans's advanced concepts, such as heterogeneous silos, versioning, and streaming.

In this chapter, we will learn about the following advanced concepts in Orleans:

- Streaming in Orleans
- Heterogeneous silos
- Grain interface versioning

By the end of this chapter, you will be able to use Orleans to build reactive applications to process streams of events in a structured way. You will also be able to leverage the versioning of grains to promote changes to production without impacting the current traffic.

Technical requirements

To follow along, you should have the following:

- Visual Studio 2022 Community Edition

- A basic understanding of building .NET applications

- The Cosmos DB Emulator from `https://aka.ms/cosmosdb-emulator`

- An Azure subscription

The code used in this chapter can be found at `https://github.com/PacktPublishing/Distributed-.NET-with-Microsoft-Orleans/tree/main/Chapter08`.

Streaming in Orleans

In typical request and response scenarios, a client sends the request to the server and the service does some processing on it and sends a response to the client. The client would be waiting for the response from the service while the service was processing it. In the event of any failures, the client will retry sending the call to the service or will do some other action such as sending an email or raising an alert.

Event stream processing is a popular paradigm for durability and performance in different kinds of applications, such as gaming, IoT, high-frequency trading, or fraud detection.

An event stream processor actively tracks and processes a stream of events in an application so that insights and risks can be identified and appropriate action can be taken. Examples of such interactions are monitoring the production line, event processing from IoT devices, smart patient care in hospitals, monitoring of traffic, and so on.

There are many technologies or tools available for stream processing. There are mainly two types:

- **Durable stream data stores** – A system that moves events from place to place efficiently so that others can access and analyze them, for example, Kafka, Azure Queue Storage, Azure Event Hubs, Azure Service Bus, and so on.

- **Stream compute systems** – A system that processes the stream of events, for example, Apache Spark Streaming, Azure Stream Analytics, Apache Storm, and so on.

The systems mentioned above process large volumes of similar events in a uniform way. They do the same set of transformations, aggregations, and filtering operations on the stream of events. There is limited support, or no support, for context-aware processing of events. Orleans streams are built to bridge that gap. Orleans streams support fine-grained free-form compute over streams of data. Let's understand what that means with an example. Say we have a stream of events going to our Distel users regarding some activities that are happening, where some users are interested in sports or adventure events, and some are interested in cultural events. We want the event processing to happen in the context of the individual user. The processing logic should be dynamic, it should change when there is a change in the user's interest. The technologies we discussed above may not support such scenarios. Streams in Orleans are built to support scenarios like these. Like grains, streams are also virtual. They are present all the time.

In the Distel hotel chain, consider that there is a display board at the entrance of the hotels showing nearby upcoming attractions or events along with the dates. The events that are displayed on the board were also sent to checked-in users, perhaps via SMS or push notifications. To build this feature, at the time of guest check-in, we can subscribe the user to the notifications that the hotel publishes about any events. Let's use Orleans streams to build this feature by following these steps:

1. First, configure `Distel.WebHost` to send stream data. To do this, update the top-level statements in `Program.cs` as shown in the following code snippet:

```
builder.Host.UseOrleans(siloBuilder =>
{
    siloBuilder
    .UseLocalhostClustering()
    .AddMemoryGrainStorageAsDefault()
    .AddSimpleMessageStreamProvider("attractions-
        stream")
    .AddMemoryGrainStorage("PubSubStore")
    .UseInMemoryReminderService()
    .Configure<ClusterOptions>(opts =>
    {
        opts.ClusterId = "dev";
        opts.ServiceId = "DistelAPI";
    })
```

```
        .Configure<EndpointOptions>(opts =>
        {
            opts.AdvertisedIPAddress = IPAddress.Loopback;
        });
    });
```

In the preceding code snippet, we are registering `SimpleMessageStreamProvider` (SMS) from Orleans with the name `"attractions-stream"` by calling the `AddSimpleMessageStreamProvider` method. `SimpleMessageStreamProvider` delivers the events to subscribers by using the Orleans grain messaging system over TCP. To store the list of subscriptions, we are using the in-memory storage provider, which is configured by calling the `AddMemoryGrainStorage` method. The storage provider used by Orleans streams should be registered with the reserved-name `PubSubStore`. In-memory storage providers should not be used in production scenarios; we should be using the persistence storage providers that we discussed in *Chapter 5, Persistence in Grains*. With this configuration, our host is ready to support Orleans streams.

2. Let's build the functionality of publishing events in `HotelGrain`. For this, add the `PublishEvent` method to the `IHotelGrain` interface as shown here:

```
public interface IHotelGrain : IGrainWithStringKey
{
    //Code removed for brevity
    Task PublishEvent(AttractionNotification
        attractionNotification);
}
```

In the preceding code snippet, we added the `PublishEvent` contract to the `IHotelGrain` interface, which takes `AttractionNotification` as an argument.

3. Add the `AttractionNotification` model class to the `Distel.Grains.Interfaces` project as shown here:

```
public class AttractionNotification
{
    public string Description { get; set; }
    public DateTime EventDate { get; set; }
    public override string ToString() =>
        $"There is \"{Description}\" on {EventDate}";
```

This model defines the two attributes that denote the description and date of the event. We are also overwriting the ToString method, which we will be using just for logging purposes in this demo.

4. Now let's add the functionality of publishing the events to HotelGrain. On grain activation, register the stream provider and use it to publish the event:

```
public class HotelGrain : Grain, IHotelGrain
{
//Code removed for Brevity
    private IAsyncStream<AttractionNotification>
        stream;
    private Guid displayBoardId = Guid.NewGuid();
    public override Task OnActivateAsync()
    {
        this.stream = base
            .GetStreamProvider("attractions-
                stream")
            .GetStream<AttractionNotification>(
                displayBoardId,
                "AttractionEvents-NS");
        return base.OnActivateAsync();
    }
    public async Task
    PublishEvent(AttractionNotification
        attractionNotification) =>
        await this.stream
            .OnNextAsync(attractionNotification);
}
```

In the preceding code snippet, we are getting the stream provider with the name that is registered at the silo host configuration by calling GetStreamProvider. From the retrieved stream provider, we are getting the stream by calling the GetStream method by passing the displayboardId argument, which is a unique key for each hotel, and putting "AttractionEvents-NS" as stream namespace. Now, when there is a call to the PublishEvent method, we will simply publish the event to stream by calling the OnNextAsync method by passing in the attraction notification.

5. Let's now build the functionality in `UserGrain` to subscribe to the events published by `HotelGrain`. For this, define the `SubscribeToAttractionsEventsAsync` method. which will be called to let the `User` grain subscribe to events. Update the `IUserGrain` interface as shown in the following code snippet:

```
public interface IUserGrain : IGrainWithStringKey
{
    Task<TravelHistory> GetTravelHistory();
    Task SubscribeToAttractionEventsAsync(Guid
        displayBoardId, string nameSpace);
}
```

In the preceding code snippet, `SubscribeToAttractionsEventsAsync` takes two parameters. The first parameter is the ID of the display board and the second parameter is the namespace that is used in subscribing to the stream.

6. Add the functionality to `UserGrain` to subscribe to the events as shown in the following code snippet:

```
public class UserGrain : Grain, IUserGrain
{
// Code removed for brevity
    public Task SubscribeToAttractionEventsAsync(Guid
        displayBoardId, string nameSpace) =>
        GetStreamProvider("attractions-stream")
        .GetStream<AttractionNotification>
        (displayBoardId, nameSpace)
        .SubscribeAsync(new
        AttractionObserver(Notifier));

    private Task Notifier(AttractionNotification
        notification)
    {
        this.logger.LogInformation
            (notification.ToString());
        return Task.CompletedTask;
    }
}
```

In the implementation of `SubscribeToAttractionEventsAsync`, we are getting the stream provider with the same name registered with the `"attractions-stream"` silo and retrieving the stream with `Key` and the `namespace` passed in as arguments. We subscribed to the stream by calling the `SubscribeAsync` method on the stream by passing in the `AttractionObserver` observer. Let's implement `AttractionObserver` in the
next step.

7. The subscriber to the stream in Orleans should implement the `IAsyncObserver` interface. This interface is similar to the .NET `IObserver` interface. All the consumers of the Orleans streams should implement the `IAsyncObserver` interface. Add the call implementation of `AttractionObserver` as shown in the following code snippet:

```
class AttractionObserver :
  IAsyncObserver<AttractionNotification>
{
private readonly Func<AttractionNotification, Task>
  action;
public AttractionObserver(Func<AttractionNotification,
  Task> action) =>
    this.action = action;
public Task OnCompletedAsync() => Task.CompletedTask;
public Task OnErrorAsync(Exception ex) =>
  Task.CompletedTask;
public Task OnNextAsync(AttractionNotification item,
  StreamSequenceToken token = null) =>
          action(item);
}
```

In the preceding code snippet, `AttractionObserver` implements the generic `IAsyncObserver` interface to consume the stream of events. This interface has three methods:

*   `OnNextAsync`: The next item is passed to the consumer through this method. This method should return `Completed Task` post-processing.

*   `OnErrorAsync`: This method is called to notify the consumer that the stream had an error. The consumer should return `Completed Task` when the processing is complete.

- OnCompletedAsync: This method is called to notify the consumer that the stream is closed. The task returned from this method should be Completed when the consumer is done processing the stream closure.

OnErrorAsync and OnCompletedAsync are action methods to notify the consumer to take appropriate action, they do not mean the stream is closed or deactivated. Streams in Orleans always exists virtually.

In our sample, we have the implementation only for the OnNextAsync method to call the action passed into the ActorObserver class. This simply logs the event received.

8. Now let's make UserGrain subscribe to the events published by HotelGrain at the time of check-in from HotelGrain. Modify the CheckInGuest method of HotelGrain as shown in the following code:

```
public async Task<string> CheckInGuest(UserCheckIn
   userCheckIn)
{
    // TODO: Build allotment component
    checkedInGuests.State.Add(userCheckIn);
    await this.checkedInGuests.WriteStateAsync();
    var userGrain =
      this.client.GetGrain<IUserGrain>
      (userCheckIn.UserId);
    await userGrain
    .SubscribeToAttractionEventsAsync(displayBoardId,
    "AttractionEvents-NS");
    return "";
}
```

In the preceding code, we are getting the reference to UserGrain and calling the SubscribeToAttractionEventsAsync methods with the display board ID and "AttractionEvents-NS" as the namespace. This will make the user grain subscribe to the events that HotelGrain is going to publish on the display board. Now we have all the implementation required to publish and consume the streams in Orleans.

9. Before executing the code, let's add an action method to `DistelController` to simulate `HotelGrain` receiving the attraction notifications. Add `PublishAttractionEvent` as shown in the following code snippet. This will publish `AttractionNotification` received by the respective `HotelGrain` class:

```
[HttpPost("/{hotel}/event")]
public async Task<IActionResult>
 PublishAttractionEvent ([FromRoute] string hotel,
  [FromBody] AttractionNotification attraction)
{
    var hotelGrain =
      this.clusterClient.GetGrain<IHotelGrain>(hotel);
    await hotelGrain.PublishEvent(attraction);
    return Ok();
}
```

In the preceding code, we are getting the reference to `HotelGrain` from the Orleans runtime and calling `PublishEvent` with the event received as an argument. This will publish the event via the stream we registered in *Step 4*.

10. Let's run the application and execute the `CheckIn` action from Swagger UI, which will make `User` grain register for the stream. Now run the `PublishEvent` action method. This will make `HotelGrain` publish the event. The subscribed `UserGrain` receives the event. We can see this in the console log as shown in the following screenshot:

```
2
warn: Orleans.Runtime.AsyncTimer.IncomingRequestMonitor[0]
      Timer should have fired at 10/30/2021 16:30:20 but fired at 10/30/2021 16:30:25, which is 00:00:05.0210972 longer
than expected
info: Distel.Grains.UserGrain[0]
      There is "Concert by Taj rock band" on 10/30/2021 4:27:40 PM
warn: Orleans.Runtime.Scheduler.WorkItemGroup[101215]
      Task [Id=2084, Status=RanToCompletion] in WorkGroup [Activation: S127.0.0.1:11111:373307328*grn/Distel.Grains.User
Grain/0+string@fa6c3487 #GrainType=Distel.Grains.UserGrain Placement=RandomPlacement State=Valid] took elapsed time 0:00
:07.106841 for execution, which is longer than 00:00:00.2000000. Running on thread 12
```

Figure 8.1 – Explicit subscription grain output

With this, we have built the feature to publish and consume streams of attractions in our Distel application using Orleans streams. With this example, what we have seen is an explicit subscription by `UserGrain` to a stream published by `HotelGrain`. In the next section, let's see how to configure a grain to subscribe to a stream implicitly.

Implicit subscriptions

In the previous example, the User grain is explicitly subscribing to the stream by calling the SubscribeAsync method on the stream. **Explicit subscription** means the grain retrieves the stream handle and subscribes to it by calling the SubscribeAsync method as we saw in the previous section. Orleans also supports **implicit subscriptions** to streams. In this section, let's learn how to use implicit subscriptions. With implicit subscriptions, the Orleans runtime subscribes the grain to the specified stream without the grain calling the SubscribeAsync method. The display board that we have at the front office should be subscribing to the events and displaying them. Let's define DisplayBoardGrain, which subscribes to the stream implicitly as shown here:

```
[ImplicitStreamSubscription("AttractionEvents")]
public class DisplayBoardGrain : Grain, IDisplayBoardGrain,
    IStreamSubscriptionObserver
{
    private readonly ILogger<DisplayBoardGrain> logger;
    private readonly AttractionObserver observer;
    public DisplayBoardGrain(ILogger<DisplayBoardGrain>
        logger)
    {
        this.logger = logger;
        this.observer = new AttractionObserver(this.Notifier);
    }
}
```

The ImplicitStreamSubscription attribute takes the namespace of the stream as a parameter. The Orleans runtime subscribes the grain to the specified stream namespace.

The grain also implements the IStreamSubscriptionObserver interface with the OnSubscribed method. The OnSubscribed method called by the Orleans runtime to notify the grain that it is subscribed to the configured stream. When the grain is subscribed to the stream, we will attach the AttractionObserver obeserver, which we created in the previous section, to process the messages received. In the constructor, we created the instance of AttractionObserver, which we will be using to subscribe to the events from the stream.

Now implement the `IStreamSubscriptionObserver` interface as shown in the following code snippet:

```
[ImplicitStreamSubscription("AttractionEvents")]
public class DisplayBoardGrain : Grain, IDisplayBoardGrain,
IStreamSubscriptionObserver
{
    // Code removed for brevity
  public async Task
  OnSubscribed(IStreamSubscriptionHandleFactory
    handleFactory)
    {
        var handle =
          handleFactory.Create<AttractionNotification>();
        await handle.ResumeAsync(this.observer);
    }
    private Task Notifier(AttractionNotification
      notification)
    {
        this.logger.LogInformation("DisplayBoardGrain: " +
          this.GetPrimaryKey().ToString() +
          notification.ToString());
        return Task.CompletedTask;
    }
}
```

In the preceding code, the `OnSubscribed` method is implemented to create the handler for the stream by calling the `Create` method on `IStreamSubscriptionHandleFactory` for the message type `AttractionNotification`. Attach the `AttractionObserver` observer that we created in the constructor by calling `ResumeAsync` on the subscription handler.

When a message is received, it will be routed to the `Notifier` method, which will just log the message received in our Distel demo application.

Let's run the application and execute the actions in Swagger UI as we did before. First, execute user check-in and then call /{hotel}/event. In the output console, we see that the published event is received by the explicitly subscribed UserGrain class and implicitly subscribed DisplayBord grain as shown in the following screenshot.

```
2
warn: Orleans.Runtime.Scheduler.WorkItemGroup[101215]
      Task [Id=831, Status=RanToCompletion] in WorkGroup [Activation: S127.0.0.1:11111:373393214*grn/Distel.Grains.Hotel
Grain/0+Taj@787aa6b3 #GrainType=Distel.Grains.HotelGrain Placement=RandomPlacement State=Valid] took elapsed time 0:00:0
1.2128325 for execution, which is longer than 00:00:00.2000000. Running on thread 14
info: Distel.Grains.UserGrain[0]
      UserGrain: There is "Live concert by rock band" on 10/31/2021 3:48:34 PM
warn: Orleans.Runtime.Scheduler.WorkItemGroup[101215]
      Task [Id=906, Status=RanToCompletion] in WorkGroup [Activation: S127.0.0.1:11111:373393214*grn/Distel.Grains.UserG
rain/0+string@afade38e #GrainType=Distel.Grains.UserGrain Placement=RandomPlacement State=Valid] took elapsed time 0:00:
03.0021606 for execution, which is longer than 00:00:00.2000000. Running on thread 14
info: Distel.Grains.DisplayBoardGrain[0]
      DisplayBoardGrain: There is "Live concert by rock band" on 10/31/2021 3:48:34 PM
warn: Orleans.Runtime.Scheduler.WorkItemGroup[101215]
      Task [Id=936, Status=RanToCompletion] in WorkGroup [Activation: S127.0.0.1:11111:373393214*grn/Distel.Grains.Displ
ayBoardGrain/f5268c55-74fa-4fa1-b896-b1d9554c7c14@1522ef80 #GrainType=Distel.Grains.DisplayBoardGrain Placement=RandomPl
acement State=Valid] took elapsed time 0:00:02.8642334 for execution, which is longer than 00:00:00.2000000. Running on
thread 5
```

Figure 8.2 – Implicit subscription grain output

Here, when the message is published by HotelGrain, the Orleans runtime will activate DisplayBoardGrain with the stream ID used to get the stream as the primary key. In this case, we used displayBoardId that we created at the time of HotelGrain activation. For the demo, we are creating a new ID with every grain activation. Ideally, it should be the ID of the display board and it should not change with every activation. We may be persisting it in the HotelGrain state.

To learn about the concepts of streaming in Orleans, we used the Simple Message Stream Provider. This does not guarantee reliable event delivery and it does not resend the message automatically in the event of failures.

As we saw with state persistence in *Chapter 5*, *Persistence in Grains*, Orleans follows the provider model. For streams, there are also multiple providers available, such as the Azure Queue Storage provider and the Azure Event Hubs provider. Orleans also provides the generic extensible PersistentStreamProvider, which enables us to plug in different types of queues without building everything from scratch. The queue processing is made extensible through the IQueueAdapter interface in PersistentStreamProvider. The Azure Event Hubs provider and Azure Queue Storage provider are implemented in the same way. To use the Event Hubs stream provider, we will be configuring the silo as shown in the following code snippet:

```
hostBuilder
  .AddAzureQueueStreams("AzureQueueProvider", configurator
  => {
    configurator.ConfigureAzureQueue(
```

```
      ob => ob.Configure(options => {
        options.ConnectionString = <<Azure queue
          connectionstring>>;
        options.QueueNames = new List<string> {
          "queuename1" };
      }));
  })
.AddAzureTableGrainStorage("PubSubStore", options => {
    options.ConnectionString = <<Azure table connection
      string>>;
  })
```

In the preceding code snippet, the Azure Queue Storage stream provider is configured with the required details, the Azure Queue Storage connection string, and the queue name.

Here are a few things that we should be aware of regarding Orleans streams:

- If a consumer subscribes to a stream N times, the message will be delivered to that consumer N times.

- To find all the subscriptions, the consumer can call the GetAllSubscriptionHandles method as shown here:

```
Var allSubscriptions = await
  GetStreamProvider("attractions-stream")
.GetStream<AttractionNotification>(displayBoardId,
  nameSpace)
.GetAllSubscriptionHandles()
```

- If the consumer of the stream dies or gets de-activated, it will be activated again, and whenever there is a new message, it is sent to the consumer grain.

- If the producer of the messages dies, the producer will get the handle of the stream next time when it needs to send the messages.

- All those subscribers who subscribed to the stream will receive the message published by a producer. Those subscribers who subscribed after the message was published will not receive the old messages; they will receive the events published after they subscribed.

In this section, we have learned how Orleans provides reliable, scalable stream processing with elasticity. In the next section, let's learn about another important aspect: heterogeneous silos.

Looking into heterogeneous silos

We deploy an Orleans application like **Distel.Host** on a node that is called a silo. Multiple silos form a cluster. The nodes in the cluster share the load and grain activations. Our understanding so far is, in Orleans, a given grain can be activated on any of the nodes in the cluster to make use of the resources uniformly. These grain activations will be maintained in the grain directory. What if we wanted a specific type of grain activated on a specific node?

Consider a situation where a grain requires some special resources, or perhaps some specific hardware. It does not make sense to install that hardware in all the nodes in the cluster. Or, we want the calls to an external service from a grain to be restricted to go out from a specific node. Or, we want to split our application into microservices where a node hosts only the defined grains so that we can independently scale resources required for a grain type. This is supported by Orleans with heterogeneous silos.

Consider the following diagram, which shows the silos with the supported grains of our Distel application.

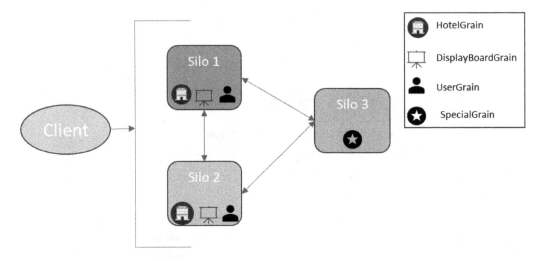

Figure 8.3 – Heterogeneous silo

In the preceding diagram, the cluster has three nodes. **Silo 1** and **Silo 2** host `HotelGrain`, `DisplayBoardGrain`, and `UserGrain`. Silo 3 only hosts `SpecialGrain`. When there is a request for `SpecialGrain` from the client, the Orleans runtime activates `SpecialGrain` on **Silo 3**. There is no special configuration required for heterogeneous silos. Every silo in the cluster should know about all the grain interfaces. In the above example, **Silo 1** and **Silo 2** also know about the `ISpecialGrain` interface. The grain implementation `SpecialGrain` will only be present in **Silo 3**. When there is a request from a client for `ISpecialGrain`, it will be routed to **Silo 3**. Let's build this heterogeneous silo in our Distel application by following these steps:

1. Add the grain interface `ISpecialGrain` to the `Distel.Grains.Interfaces` project:

    ```
    public interface ISpecialGrain : IGrainWithGuidKey
    {
        Task<string> SpecialAction();
    }
    ```

2. Create a new project, `Distel.Grains2`, similar to the `Distel.Grains` project we built in *Chapter 4, Understanding Grains and Silos*, and implement `SpecialGrain` as shown here:

    ```
    public class SpecialGrain : Grain, ISpecialGrain
    {
        public Task<string> SpecialAction()
        {
            return Task.FromResult("From special grain");
        }
    }
    ```

3. Our existing host, `Distel.WebHost`, is currently hosting the `UserGrain`, `HotelGrain`, and `DisplayBoardGrain` grains, which we have built so far.

    We will build a new silo, `Distel.WebHost2`, which will only host `SpecialGrain`. Our intention is to route all the requests that come to `SpecialGrain` to the new silo, `Distel.WebHost2`. Build the `Distel.WebHost2` host project like we built `Distel.WebHost` in *Chapter 2, Cloud Architecture and Patterns for Distributed Applications*. To this project, add a reference to the `Distel.Grains.Interfaces` and `Distel.Grains2` projects.

4. We will be running both the silos from the same developer machine. So, we should run these silos with a different silo and gateway port. Configure the `Distel.WebHost2` silo in the `Program` class as shown in the following code snippet:

```
// Code removed for brevity
builder.Host.UseOrleans(siloBuilder =>
{
--------------------------
    siloBuilder
    .ConfigureEndpoints(IPAddress.Loopback, siloPort:
      11112, gatewayPort: 30001,
      listenOnAnyHostAddress: true)
    .UseAzureStorageClustering((options) =>
    {
        options.TableName = "distelcluster";
        options.ConnectionString = <<Azure table
        storage connection string>>;
    })
--------------------------
});
```

In the preceding code snippet, we are running the silo on `siloPort` `11112` and `gatewayport` `30001`. We are using Azure Table Storage for clustering. Add the package reference to `Microsoft.Orleans.Clustering.AzureStorage` to use Azure Table Storage clustering.

5. Modify the `Distel.WebHost` silo configuration also to use Azure Table Storage the same as the code snippet in step 4, but with `siloport` as `11111` and `gatewayport` as `30000`:

```
.ConfigureEndpoints(IPAddress.Loopback, siloPort: 11111,
gatewayPort: 30000, listenOnAnyHostAddress: true)
```

6. Now add the `Specialaction` action method `DistelController` in both host projects, `Distel.WebHost` and `Distel.WebHost2`, as shown here:

```
[HttpGet("/specialaction")]
public async Task<IActionResult> SpecialAction()
{
```

```
      var hotelGrain =
        this.clusterClient.GetGrain<ISpecialGrain>
        (Guid.NewGuid());
      var message = await hotelGrain.SpecialAction();
      return Ok(message);
  }
```

In the preceding code snippet, we are getting the reference to `ISpecialGrain` and calling the `SpecialAction` method on it.

7. Now set `Distel.WebHost` and `Distel.WebHost2` as startup projects and run the solution. This will launch both the hosts and render Swagger UI in the browser. You execute `SpecialAction` from Swagger UI. You get the same response from both hosts as shown in the following screenshot.

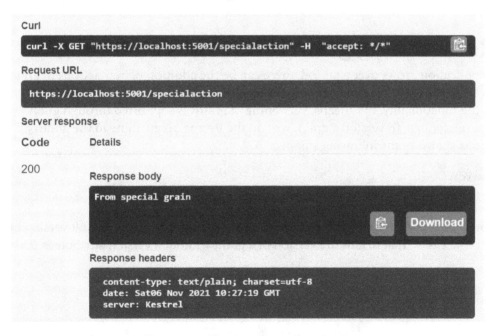

Figure 8.4 – Heterogeneous silo response

From Swagger UI for the host `Distel.WebHost2`, you will also be able to execute operations on `HotelGrain` or `UserGrain`, though it does not have a reference to the `Distel.Grain` class. Here, the `Distel.WebHost` and `Distel.WebHost2` silos are forming a cluster. The Orleans runtime will route the requests to the respective silos that have the grain implementations.

Though heterogeneous silos are a great way to address certain situations, which we talked about earlier, there are a few limitations to them:

- Stateless grains are not supported in heterogeneous silos; all the silos in a cluster must support all stateless grains.

- Implicit streaming subscription grains are not supported; only explicit streaming subscription grains are supported.

- If a failure occurs and the silo that hosts a certain grain type departs the cluster, the client's request will still be routed to that silo, and the client will receive an exception of type `OrleansException` since no other silo hosts that grain type.

In this section, we have learned how Orleans supports heterogeneous silos. In the next section, we will learn about one of the most important aspects of large distributed applications, that is, interface versioning.

Understanding grain interface versioning

As the application grows over a period, there can be the addition of new actions to grains or changes to the existing implementations. To adhere to these changes, we normally version the functionality. In Orleans, versioning of grains is supported through grain interface versioning. To version a grain, we add the `Version` attribute to the grain interface as shown in the following snippet:

```
[Version(1)]
public interface IHotelGrain : IGrainWithStringKey
```

In the preceding code snippet, the version of `IHotelGrain` is 1. The default version of a grain interface is 0. That means the version of a grain without a version attribute is 0.

Grain activation with versions

The following flow chart depicts the activation of a versioned grain:

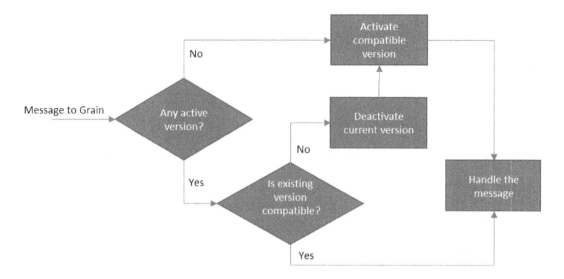

Figure 8.5 – Version grain activation flow

When there is a message to a versioned grain, the Orleans runtime first checks whether there is an existing activation. If there is no existing activation, the runtime will activate a compatible grain. In the case of an existing activation, the runtime will check whether it is compatible. If compatible, the request will be processed by the existing activation. If it is not compatible, the existing activation will be deactivated, and a compatible version of the grain will be activated to process the message.

The compatibility strategy will be determined by DefaultCompatibilityStrategy. The compatibility strategy will be set at the time of silo initialization. The strategy cannot be changed at runtime. Following are the different compatibility strategies supported by Orleans:

- StrictVersionCompatible: Strict version compatibility, that is, there is no compatibility. Only the requested grain version can serve the request.

- BackwardCompatible: Versions are backward compatible. For example, version 2 activation can serve the requests for version 1, but version 1 activation cannot serve version 2 requests.

- AllVersionsCompatible: All versions are compatible with all versions. A request can be served by any version of the grain.

Version selection of the grain will be determined by `DefaultVersionSelectorStrategy`. A version selection strategy should also be set at the time of silo initialization and cannot be changed at runtime. Following are the different version selection strategies supported by Orleans:

- `LatestVersion`: Select the latest version from the compatible versions.
- `MinimumVersion`: Select the minimum version from the compatible versions.
- `AllCompatibleVersions`: Select all compatible versions.

The compatibility strategy and version selection strategy will form the `GrainVersioningOption` configuration, which can be set during the silo's configuration as shown in the following code snippet:

```
var builder = new SiloHostBuilder()
  .Configure<GrainVersioningOptions>(options =>
  {
    options.DefaultCompatibilityStrategy =
      nameof(BackwardCompatible);
    options.DefaultVersionSelectorStrategy =
      nameof(MinimumVersion);
  })
  [...]
```

In the preceding code snippet, `DefaultCompatibilityStrategy` is set to `BackwardCompatible` and `DefaultVersionSelectorStrategy` is set to `MinimumVersion`.

> **Note**
> Refer to the Orleans documentation for the backward compatibility guidelines: `https://dotnet.github.io/orleans/docs/grains/grain_versioning/backward_compatibility_guidelines.html`.

Now we know about the different compatibility and version section strategies supported by Orleans, in the next section, we will see some recommendations for these based on the deployment practice.

How to deploy new versions of grains

Most of the time, we strive for zero-downtime deployments to push a change to production. Typically, the methods followed at the time of deployment are rolling upgrades and using a staging environment. Here are the strategy recommendations from Orleans for version selection:

- **Rolling upgrades**: A node will be taken out of rotation to deploy the changes and then put back into rotation. Until the deployment is complete on all the machines, there will be a mix of nodes with the old and new versions of grains. The recommended configuration for rolling upgrades is that `DefaultCompatibilityStrategy` should be set to `BackwardCompatible`. `DefaultVersionSelectorStrategy` should be set to `AllCompatibleVersions`.

  With this configuration, the clients with the old grain version can talk to the actions on both versions of silos. Clients with the new grain version will trigger activations on newer silos.

- **Using a staging environment**: In this case, there will be two environments, production and staging. The production environment will take the live traffic. The staging environment is a production-like environment but will not take live traffic. Here, the deployment is done first in a staging environment. Once the deployment is complete in the staging environment, a percentage of production traffic will be routed to the staging environment. If everything looks fine, the staging environment will be swapped to make it a new production environment. The recommended configuration of `GrainVersioningOptions` for this is that `DefaultCompatibilityStrategy` should be set to `BackwardCompatible`. `DefaultVersionSelectorStrategy` should be set to `MinimumVersion`.

  Once the staging environment starts taking the traffic, the V2 clients and V2 silos will start creating the new version of grains in the staging nodes. If we notice any failures when the traffic is routed to the old environment, we can simply turn off the staging nodes. V2 activations will be destroyed and V1 activations will be created as needed. If everything goes fine, we will route 100% of the traffic to staging nodes. The latest version of activations will be created in new silos and, eventually, the V1 activations will be migrated to V2 versions.

In this section, we have learned about grain interface versioning and strategies to pick the right version of a grain. Something to note about grain versions is stateless worker grains and streaming interfaces do not support versioning.

Summary

In this chapter, we were introduced to some of the advanced concepts in Orleans. We learned about Orleans streams and implicit and explicit subscriptions to these streams. Using streams, we saw how our Distel application processes a stream of messages. We also learned how we can make a specific grain activated on a predefined silo using heterogeneous silos. If there is a need to scale grains independently, with heterogeneous silos, we can split our application into microservices where a node hosts only the defined grains, which makes them scale independently.

In this chapter, we also explored grain interface versioning and different strategies for selecting a versioned grain. We also looked into a few recommendations for choosing these strategies.

So far, we have learned about different concepts in Orleans. In the next chapter, let's learn about design patterns in Orleans, which represent the best practices used by experienced Orleans developers.

Questions

1. Which interface is implemented by Orleans stream consumers?

    A. IAsyncObserver

    B. IObservableAsync

    C. IOrserver

    D. IObservable

    Answer – A

2. A grain can subscribe to a stream only once.

    A. True

    B. False

    Answer – B

3. Stateless grains should be supported by all silos in a cluster.

    A. True

    B. False

    Answer – A

4. Stateless grains can be versioned with grain interface versioning.

   A. True

   B. False

   Answer – B

5. What is the recommended `DefaultVersionSelectorStrategy` for the `GrainVersioningOptions` configuration to perform a rolling upgrade?

   A. `AllCompatibleVersions`

   B. `LatestVersion`

   C. `MinimumVersion`

   D. None

   Answer – A

9
Design Patterns in Orleans

In the last chapter, you learned about a few advanced concepts in Orleans, such as streaming and heterogeneous silos. You may not use them in every application but they will likely be used when you really get into the ecosystem. Design patterns represent the best practices used by experienced developers in a given field.

A **design pattern** is a generic, reusable solution to a common software design problem that we face during development. It's not something you'll utilize right away in your application. It is a template for how to solve a problem that can be used in many different solutions. You might have referenced design patterns in various situations. The implementations of design patterns may vary with language and technology.

In this chapter, we will learn about some of the design patterns published by Orleans contributors:

- Distributed cache
- Batch message processing with Dispatcher
- Cadence with timers
- Aggregating with the Reduce pattern

By the end of this chapter, you will be able to use some of the design patterns that are popular in the Orleans community.

Technical requirements

To follow along with the chapter, you need to have the following:

- Visual Studio 2022 Community Edition
- A basic understanding of building .NET applications

The code used in this chapter can be found at:

`https://github.com/PacktPublishing/Distributed-.NET-with-Microsoft-Orleans/tree/main/Chapter09`.

Distributed cache

As we all know, information technology has dramatically improved business processes over the years and has become crucial for businesses all over the world. Applications are at the heart of any organization, and their use has increased dramatically in recent years. **Data retrieval time** is a vital requirement in almost all applications and plays an important part in the user experience. If we fetch the data from the database for every request, the application will be sluggish.

A **cache** is a memory buffer that temporarily stores frequently accessed data. Because data does not have to be fetched from the original source, performance is improved. Caching is extremely important to increase data throughput. You might have already used different cache systems in various applications. In this section, we will learn how to build a cache system using **Microsoft Orleans**.

We learned from *Chapter 5, Persistence in Grains*, that upon activation, a grain loads the state from a persistent data store to memory and we call the `WriteStateAsync` method to persist the state to the database. For an Orleans application, we do not need a cache system as the data is always from the in-memory state of the grain. We will leverage this to build a distributed cache system using Orleans. We are going to build `CacheGrain`, which provides the distributed cache functionality to `ClientSystem`, as shown in the following figure:

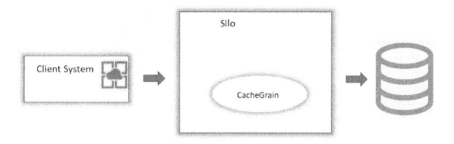

Figure 9.1 – Distributed cache pattern

Let's follow these steps to build a distributed cache system:

1. Add the `ICacheGrain` interface as shown in the following code snippet:

```
public interface ICacheGrain<T> : IGrainWithStringKey
{
    Task SetAsync(Immutable<T> value);
    Task<Immutable<T>> GetAsync();
}
```

The `ICacheGrain` interface defines the `SetAsync` and `GetAsync` methods to write and read from the cache. Since this grain is not going to change the passed data, we marked the data that we persist in the grain as `Immutable`. As we learned in *Chapter 3, Introduction to Microsoft Orleans*, marking an `Immutable` type tells the Orleans runtime to pass the data without deep copying to optimize the serialization.

2. Let's now implement the grain as shown in the following snippet:

```
public class CacheGrain<T> : Grain<Immutable<T>>,
    ICacheGrain<T>
{
    public Task<Immutable<T>> GetAsync()   =>
        Task.FromResult(State);

    public async Task SetAsync(Immutable<T> value)
    {
        State = value;
        await base.WriteStateAsync();
    }
}
```

The preceding code snippet shows the simple implementation of `CacheGrain`. The `GetAsync` method returns the state of `Grain` and the `SetAsync` method sets the passed value to the `Grain` state and calls `WriteStateAsync` to persist to the storage provider. Here, we can further optimize the `WriteStateAsync` method by delaying the write operation to the persistent state provider. In the case of write, we can only set `State` and persist the state to the provider using a timer callback.

3. Use `CacheGrain` to store and retrieve the data as shown in the following code snippet:

```
var grain =
    this.clusterClient.GetGrain<ICacheGrain<Data>>
    ("cacheKey");
var response = await grain.GetAsync();


await grain.SetAsync(new
    Orleans.Concurrency.Immutable<Data>(dataObject));
```

In the preceding code snippet, we are getting a grain activation with `cacheKey`, which is the key used to store and retrieve the data. The `GetAsync` method retrieves the cached data, and it returns `null` if there is data persisted with that key. The `SetAsync` method sets the data in the cache with the specified key.

In this sample, we have both `Get` and `Set` defined with a single `ICacheGrain` interface. We often see people defining two different interfaces and the same `CacheGrain` class implementing them to separate the read and write semantics.

We only implemented `Get` and `Set` actions in `CacheGrain`. Once the state is set in `CacheGrain`, it is available forever as we are persisting it to the grain state. Once the cache value is set, the grain will continue to serve it across activations. The user of this cache provider should think about the cache invalidation. A traditional cache provider should also have an action to clear the cache. We can implement that by simply resetting the state.

ASP.NET Core defines the contract for distributed cache providers with the `IDistributedCache` interface. You can refer to the sample code of this book on GitHub for a sample implementation of the distributed cache provider using the `CacheGrain` class we built in this section.

In this section, we have seen one of the most common patterns in Orleans, that is, distributed cache. In the next section, let's learn about another important pattern: Dispatcher.

> **Note**
> Refer to the Microsoft documentation to learn more about distributed caching in ASP.NET Core: `https://docs.microsoft.com/en-us/aspnet/core/performance/caching/distributed`

Batch message processing with Dispatcher

The **Command Dispatcher** is a well-known pattern used to isolate a command's execution from its commander. The commander doesn't need to know how the command is carried out. All that the commander needs to know is that it exists.

In IoT systems where each sensor sends lots of data for processing, it is often more effective to send messages in batches rather than sending them one at a time. There can be a dispatcher grain to distribute these messages to a specific grain. The dispatcher is in charge of coordination; it collects a batch of messages and distributes them to individual grains for processing. The Dispatcher pattern can be applied when a batch of incoming messages has to be decoded before the target grain can be determined.

As shown in the following diagram, the Dispatcher pattern has three participants:

- **Client System**: This generates the batches of messages.
- **Dispatcher Grain**: This is tasked with processing a batch of messages and disseminating them to the target grains.
- **Target Grain**: This is the grain that ultimately processes the messages.

The **Client System** sends the batches of messages to the **Dispatcher Grain**, which enumerates through the batch of messages to dispatch the messages to each target grain.

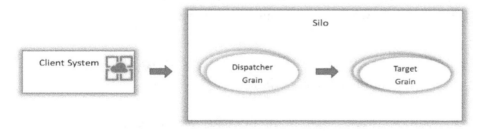

Figure 9.2 – Dispatcher pattern

Let's build a dispatcher that processes batches of feedback/comments coming to our Distel application and sends them to a specific `HotelGrain` class to process, by following the next steps:

1. Extend the `IHotelGrain` interface by defining the target grain contract to receive the message:

    ```
    public interface IHotelGrain : IGrainWithStringKey
    {
        Task ReceiveMessage(string message);
    }
    ```

2. Implement the `ReceiveMessage` interface defined by `IHotelGrain` in the `HotelGrain` class, as shown in the following code snippet:

    ```
    public Task ReceiveMessage(string message)
    {
        logger.LogInformation($"\nMessage
          received:'{message}'");
        return Task.CompletedTask;
    }
    ```

    In this sample implementation, we were just logging the received message.

3. Add a dispatcher interface, as shown in the following code snippet, to the `Distel.Grain.Interfaces` project:

    ```
    public interface IMessageDispatcherGrain :
       IGrainWithIntegerKey
    {
        Task Send(List<string> messages);
    }
    ```

4. Add the `MessageDispatcherGrain` class to the `Distel.Grain` project and implement it as shown in the following code snippet:

    ```
    [StatelessWorker]
    public class MessageDispatcherGrain : Grain,
       IMessageDispatcherGrain
    {
        public Task Send(List<string> messages)
        {
    ```

```
        var tasks = new List<Task>();
        foreach (var message in messages)
        {
            var parts = message.Split(':');
            var grain =
              GrainFactory.GetGrain<IHotelGrain>
              (parts[0]);
            tasks.Add(grain.ReceiveMessage(parts[1]));
        }
        return Task.WhenAll(tasks);
    }
}
```

The `MessageDispatcherGrain` class does not have any state associated with it. To take advantage of the stateless worker, which we learned about in *Chapter 3, Introduction to Microsoft Orleans*, we marked `MessageDispatcherGrain` as `StatelessWorker`. The `Send` method receives the list of messages from the client system as an argument. Assume that each message is prefixed with the hotel name and a colon (`:`). We are enumerating the messages received and extracting the hotel name from the message. The extracted hotel name is used to get the activation of the specific `HotelGrain` class and call `ReceiveMessage` on it to process it.

In this section, we have learned how to apply the Dispatcher pattern for batch message processing. In the next section, let's learn about the **Cadence pattern**.

Cadence with timers

As the name suggests, the **Cadence pattern** defines a rhythm to perform certain operations. For example, certain applications may want to persist the state at a defined interval rather than writing to the persistent store with every state change, but here the trade-off is latency versus acceptable staleness.

For example, in our `Distel` application, we may want to count and analyze the number of users or partners querying a specific hotel during a campaign. The participants in the Cadence pattern are *Source* (for example, `HotelGrain`) and the *Target* (for example, `CampaignGrain`) and there will be a timer registered to send updates from the **Source Grain** to the **Target Grain** at the set cadence, as shown in the following diagram:

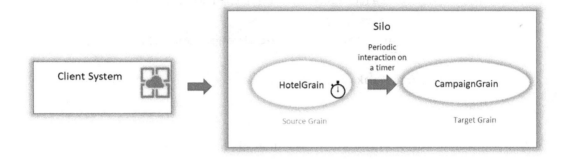

Figure 9.3 – Cadence pattern

Build the Cadence pattern by following the next steps:

1. Define the target grain interface and implement it as shown in the following code snippet:

```
public interface ICampaignGrain : IGrainWithStringKey
{
    Task ReceiveUserEngagementUpdate(int hitCount);
}
public class CampaignGrain : Grain, ICampaignGrain
{
    public Task ReceiveUserEngagementUpdate(int
      hitCount)
    {
        totalHitCount += hitCount;
        // Do something to check the campaign
            effectiveness
        return Task.CompletedTask;
    }
}
```

In the preceding code snippet, `ICampaignGrain` is the target grain interface, which defines the `ReceiveUserEngagementUpdate` method to receive the user interaction updates from a HotelGrain class.

2. Let's now update `HotelGrain`, which will be the source grain, as shown in the following code snippet:

```
public class HotelGrain : Grain, IHotelGrain
{
        public override Task OnActivateAsync()
        {
            //-------------------------------
            campaignGrain = this.GrainFactory.GetGrain
                <ICampaignGrain>("campaignKey");
            RegisterTimer(SendEngagedUpdate, null,
                TimeSpan.FromSeconds(1),
                TimeSpan.FromSeconds(10));
            //-------------------------------
            return base.OnActivateAsync();
        }
}
```

In the preceding code snippet, we are registering a timer to fire every 10 seconds when the grain activates. We are also fetching the reference to the campaign grain activation.

3. Now, we implement the timer callback, as shown in the following code snippet:

```
private async Task SendEngagedUpdate(object arg)
{
    await campaignGrain.ReceiveUserEngagementUpdate
        (hitCount);
        hitCount = 0;
}
```

When the timer callback is received, we simply update the campaign grain with the updated `hitCount`. The `hitCount` will be incremented in all those methods in `HotelGrain` that involve user interaction, such as `CheckInGuest` and `OnboardFromOtherHotel`.

We want to send the user interaction count only when the grain is active, and we don't need to send this data when the grain is not active as there won't be any change. So, we used a timer instead of a reminder. Recall from *Chapter 6, Scheduling and Notifying in Orleans*, that a timer will fire only when the grain is active, unlike a reminder, which will fire even when the grain is inactive.

In this section, we learned how to send periodic updates with the Cadence pattern. Cadence is often used with the Reduce pattern, which we will learn about in the next section.

Aggregating with the Reduce pattern

In the previous example of the Cadence pattern, we aggregated the total number of users by accessing the different Distel hotels in the `CampaignGrain` class. This will work for scenarios where the system has a small number of grains. It will start to fail with an increase in the number of grains. Moreover, there will be a performance penalty as the grains across the silos start to communicate with a **single aggregate grain** present in one of the silos.

We overcome this challenge by implementing the **Reduce pattern**, which introduces a `StatelessWorker` grain that collects results from each of the silos. This `StatelessWorker` grain then sends the number to the singleton aggregate grain, which may be in a different silo, on a regular basis.

The Reduce pattern has three participants:

- **Source Grain or Value Grain**: This is the source of the value to which aggregation is applied.

- **Subtotal or Intermediatory Grain**: This is a stateless worker grain that holds the count of user local to the silo. The source grain reports to the subtotal grain present in the same silo.

- **Aggregate or Total Grain**: This is a singleton grain that holds aggregated value. Intermediatory grains report this value to the aggregate grain at regular intervals.

The following diagram illustrates these three participants:

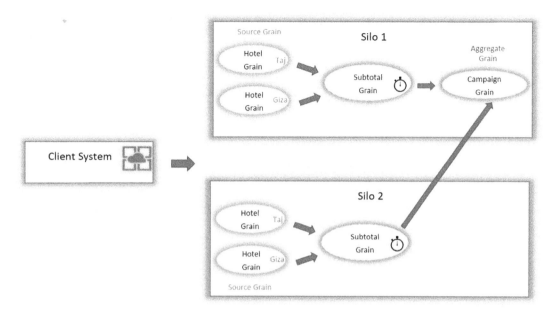

Figure 9.4 – Reduce pattern

The intermediatory grain interface can be defined as shown in the following code snippet:

```
public interface IIntermediatoryGrain :
   IGrainWithIntegerKey
{

    Task UpdateDelta(int value);

}
```

The intermediatory grain defines a method to update the delta of the count of user interactions. We implement the intermediatory grain as shown in the following code block:

```
[StatelessWorker]
public class IntermediatoryGrain : Grain,
   IIntermediatoryGrain
{

    int runningTotal = 0;
    public override Task OnActivateAsync()
    {
```

```
              RegisterTimer(ReportToAggregateGrain, null,
                TimeSpan.FromSeconds(1),
                TimeSpan.FromSeconds(10));
              return base.OnActivateAsync();
        }

    async Task ReportToAggregateGrain(object obj)
    {
       var campaignGrain =
         GrainFactory.GetGrain<ICampaignGrain>
         ("campaignKey");
         await campaignGrain.ReceiveUserEngagementUpdate
           (runningTotal);
         runningTotal = 0;
    }

    public Task UpdateDelta(int value)
    {
         runningTotal += value;
         return Task.CompletedTask;
    }
}
```

The `IntermediatoryGrain` class registers a timer that fires every 10 seconds. It maintains the running total in the `runningTotal` field and updates it to the aggregate grain, which is `CampaignGrain`, when the timer fires.

Now, we update `HotelGrain` to call `IntermediatoryGrain` instead of directly calling the `CampaignGrain` class as shown in the following code snippet:

```
private async Task SendEngagedUpdate(object arg)
{
    await intermediatoryGrain.UpdateDelta(hitCount);
    hitCount = 0;
}
```

By using the Reduce pattern, the source grain sends the updates to the intermediatory grain that is present in each silo. This call will be fast as it won't cross the process boundary. Moreover, the intermediatory grain will autoscale to the demand. This will not have any side effects as all these intermediatory grains will report to the same aggregate grain. Each intermediatory grain sends an update to the aggregate grain periodically, so the system is eventually consistent.

Summary

In this chapter, we discussed some of the design patterns widely used in applications built on Microsoft Orleans. We learned how to build a distributed cache in Orleans. You now know how to perform batch message processing with the Dispatcher pattern. We have also learned how we can use the Reduce pattern with Cadence to provide a hierarchical structure for aggregating the data stored in multiple grains.

There are more interesting patterns documented by the Orleans community. We strongly recommend referring to the Orleans contributor documentation here: `https://github.com/OrleansContrib/DesignPatterns`.

We have learned how to build a distributed application such as Distel using Microsoft Orleans. Now it is time to deploy it on the Azure cloud platform. In the next chapter, we will learn how to deploy the Orleans application in Azure Kubernetes Service.

Questions

1. While implementing the Cache pattern, the state is marked as immutable. Why?

   A. To optimize the serialization without deep copying

   B. To keep the grain state persistent

   C. No specific reason

   Answer – A

2. The dispatcher grain is built as `StatelessWorker`.

   A. True

   B. False

   Answer – A

3. The Cadence pattern is built using _____.

   A. Reminder

   B. Timer

   C. `Thread.Timer`

   D. None of the above

   Answer – B

4. The aggregate grain in the Reduce pattern is a singleton.

   A. True

   B. False

   Answer – A

Section 4 - Hosting and Deploying Orleans Applications to Azure

This section will teach you how to deploy an Orleans application to Azure Kubernetes Service and Azure App Service. We will learn how to create the necessary resources as well as the configurations needed to host in a specific environment.

In this section, we will cover the following topics:

- *Chapter 10, Deploying an Orleans Application in Azure Kubernetes*
- *Chapter 11, Deploying an Orleans Application to Azure App Service*

10

Deploying an Orleans Application in Azure Kubernetes

In the previous chapter, we learned about design patterns in Orleans such as a distributed cache, building an observer, batch message processing, and aggregation with the Reduce pattern. **Azure Kubernetes Service (AKS)** is a managed orchestration cloud service provided by Microsoft to easily deploy and manage containerized applications. Containers are widely adopted all around the globe as they facilitate microservice architecture and can be deployed, updated, and scaled independently. **AKS** has become a ubiquitous choice for containerizing your application as it eliminates the operational and maintenance burden on engineers by automatically monitoring, provisioning, and scaling resources on demand for your containerized applications. It is reliable and easy to use with an uptime SLA of 99.95% availability for Kubernetes clusters that use Availability Zones and 99.9% availability for clusters that don't use Availability Zones.

In this chapter, we will cover the following topics:

- Introduction to Azure Kubernetes
- Leveraging Azure Kubernetes hosting
- Creating a Docker image
- Pushing the Docker image to Azure Container Registry
- Deploying to Azure Kubernetes

By the end of this chapter, you will be able to deploy an Orleans application in Azure Kubernetes.

Technical requirements

To follow along with the chapter, you need to have the following:

- Visual Studio 2022 Community Edition
- An understanding of building .NET applications
- A basic understanding of how to create Azure resources
- A basic understanding of Docker and Kubernetes
- An Azure subscription

The code used in this chapter can be found at `https://github.com/PacktPublishing/Distributed-.NET-with-Microsoft-Orleans/tree/main/Chapter10`.

Understanding Azure Kubernetes

As we all know, microservice architecture consists of small, loosely coupled, independent, and autonomous services. Services can be deployed and scaled independently and communicate with each other via well-defined APIs or messaging systems such as **Azure Service Bus**. As shown in the microservice architecture in the following figure, a large application is broken down into smaller services where each service is self-contained and can be deployed and scaled independently. Each service is responsible for managing its own data stores and scenarios demanding lower latency can be optimized by bringing in a cache or high-performance NoSQL stores:

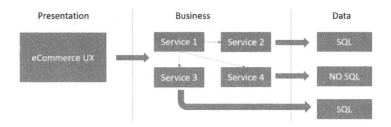

Figure 10.1 – Microservice architecture

One of the main challenges with microservices is having too many components and moving parts involved. Managing them sometimes can be very tedious, especially with no automation. This is where containers and AKS come into the picture. Containers involve a lightweight process where everything needed for an application to execute, such as code, configuration, runtime, and dependent libraries, are packaged together, and these packages can be deployed, updated, and scaled independently. **AKS** is a managed orchestration cloud service provided by Microsoft to easily deploy and manage containerized applications without the need for engineers to master the deployment and container orchestration experience.

AKS provides the following benefits:

- Better resource and hardware utilization
- Container rescheduling in the event of node failures
- Automatic scaling
- Maintenance activities such as provisioning, upgrading, and patching
- Health monitoring
- Load balancing
- An uptime SLA of 99.95% availability for Kubernetes clusters that use Availability Zones
- An uptime SLA of 99.9% availability for Kubernetes clusters that don't use Availability Zones
- Continuous integration and deployment (CI/CD)
- Security

Let's look at the steps required to run container packages in **AKS**.

Steps to run an Orleans application in Azure Kubernetes

Here is the list of steps to get your Orleans application up and running in Azure Kubernetes.

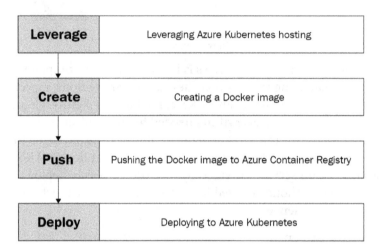

Figure 10.2 – Steps to run an application in AKS

Each of the preceding steps involves Microsoft Azure resource creation, which can be achieved in various ways:

- **The Azure CLI** (short for **Command Line Interface**) – A cross-platform command-line interface to connect to Azure and administer Azure resources.

- **Azure Resource Manager (ARM)** – Provides a management layer to manage Azure resources. The details of resources to be managed are passed in declarative JSON syntax called ARM templates.

- **Bicep** – This is a domain-specific language that simplifies resource authoring with type safety and cleaner syntax. It is a transparent abstraction over ARM templates.

- **The Azure portal** – We can also create and manage Azure resources from the Azure portal, but this is not usually done to create production resources as it's a manual process and not all engineers have access to and permissions for the portal to create resources.

Creating resources using the Azure CLI, ARM, and Bicep can be integrated with continuous integration pipelines. They all have feature parity. To create the resources for the Distel application and deploy them to Azure Kubernetes, let's use the Azure CLI and PowerShell to run the commands.

To install the Azure CLI, download the Azure CLI installer from `https://docs.microsoft.com/en-us/cli/azure/install-azure-cli` and follow the steps.

> **Note**
>
> The Azure CLI command reference list can be found here: `https://docs.microsoft.com/en-us/cli/azure/service-page/list%20a%20-%20z?view=azure-cli-latest`.

Now, let's take a closer look at the steps to run an Orleans application in Kubernetes.

Leveraging Azure Kubernetes hosting

The `Microsoft.Orleans.Hosting.Kubernetes` NuGet package provides support for hosting a Microsoft Orleans application in Azure Kubernetes. `ISiloBuilder.UseKubernetesHosting` is the extension method provided by the NuGet package. You can add this NuGet Package to the `Distel.WebHost` project and update `Program.cs` to use `siloBuilder.UseKubernetesHosting()` instead of `siloBuilder.UseLocalhostClustering()`, as shown here:

```
<<Code removed for brevity>>


builder.Host.UseOrleans(siloBuilder =>
{
// In Kubernetes, we use environment variables and the pod
    manifest
siloBuilder.UseKubernetesHosting();
});
<<Code removed for brevity>>
```

The `siloBuilder.UseKubernetesHosting()` method will take care of the following steps and hosting an Orleans application in a Kubernetes cluster:

- Setting the `SiloOptions.SiloName` property with the pod name.

- Setting the `EndpointOptions.AdvertisedIPAddress` property to the pod IP.

- `EndpointOptions.SiloListeningEndpoint` and `EndpointOptions.GatewayListeningEndpoint` will listen with the configured `SiloPort` and `GatewayPort`. Default port values of `11111` and `30000` are used if no values are set explicitly.

- Setting the `ClusterOptions.ServiceId` property with the name `orleans/serviceId`.

- Setting the `ClusterOptions.ClusterId` property with the name `orleans/clusterId`.

In the next section, we will learn how to create a Docker container image for `Distel.WebHost`.

Creating a Docker image

The Docker platform is an open source platform that helps us package applications (code, configuration, dependencies) as a standalone lightweight executable image that can then be deployed and executed in an Azure Kubernetes cluster.

> **Note**
> You can learn more about Docker and containers here:
>
> `https://docs.docker.com/get-started`
>
> `https://docs.docker.com/desktop/windows`
>
> `https://docs.docker.com/desktop/windows/install`

Through the following steps, we will see how to package our application as a Docker container image:

1. Add Docker support by right-clicking the `Distel.WebHost` project in Solution Explorer and selecting **Add** > **Docker Support...** as shown in the following figure:

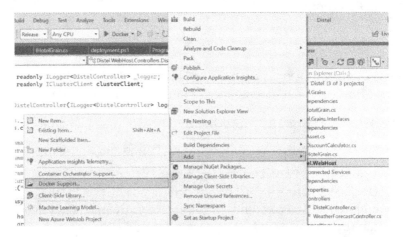

Figure 10.3 – Adding Docker support

2. Once you select **Docker Support…**, you will get an option to select the target OS as shown in the following figure. For this demo, I have selected **Linux**:

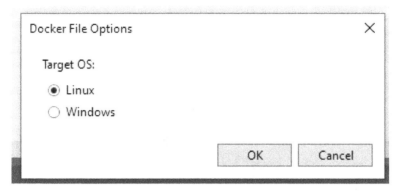

Figure 10.4 – Target OS for Docker image

3. Once you select the target OS, a Dockerfile with a set of instructions is added to the project. The following is a Dockerfile generated for the `Distel.WebHost` project. With these instructions, a Docker container image is created:

```
FROM mcr.microsoft.com/dotnet/aspnet:6.0 AS base
WORKDIR /app
EXPOSE 80
EXPOSE 443


FROM mcr.microsoft.com/dotnet/sdk:6.0 AS build
WORKDIR /src
COPY ["Distel.WebHost/Distel.WebHost.csproj",
   "Distel.WebHost/"]
COPY ["Grain.Interfaces/Distel.Grains
   .Interfaces.csproj",
   "Grain.Interfaces/"]
COPY ["Grains/Distel.Grains.csproj", "Grains/"]
RUN dotnet restore
   "Distel.WebHost/Distel.WebHost.csproj"
COPY . .
WORKDIR "/src/Distel.WebHost"
RUN dotnet build "Distel.WebHost.csproj"
   -c Release -o /app/build
```

```
FROM build AS publish
RUN dotnet publish "Distel.WebHost.csproj"
  -c Release -o /app/publish

FROM base AS final
WORKDIR /app
COPY --from=publish /app/publish .
ENTRYPOINT ["dotnet", "Distel.WebHost.dll"]
```

4. Build a Docker image from the Dockerfile using the following command:

```
docker build
  -f "PROJECT_PATH\Dockerfile"
  -t IMAGE_NAME:dev
  "SOLUTION_PATH"
```

Below is the command I used for this demo. Replace "D:\Orleans\Chapter 10\Code\ Distel" with your folder name where Distel.WebHost resides:

```
docker build -f "D:\Orleans\Chapter 10\Code\
  Distel\Distel.WebHost\Dockerfile" -t
  distelwebhost:development "D:\Orleans\Chapter
  10\Code \Distel"
```

5. Once the command executes successfully, you can open Docker Desktop to check whether the image has been created locally.

Figure 10.5 – Docker Desktop

In this section, we have created a Docker image for our `Distel.WebHost` application in the local repository. In the next section, let's push this image to **Azure Container Registry (ACR)**.

Pushing the Docker image to Azure Container Registry

ACR is offered by Microsoft to securely store and manage container images. It provides integrated security with **Azure Active Directory** authentication and role-based access control. It can connect to Azure Kubernetes Service and **Azure App Service** environments, which help in enabling **continuous integration and deployment (CI/CD)**. Let's see how to provision Azure resources and create a container registry that can be used to store our Docker container image created in the previous section:

> **Note**
>
> You can learn about ACR in detail here: `https://azure.microsoft.com/en-in/services/container-registry/#overview`.

1. Define the variables and values:

```
$resourceGroup = "distelrg"
$location = "westus"
$storageAccount = "distelstorage"
$clusterName = "distelwebhost"
$containerRegistry = "distelwebhostacr"
```

2. Log in to the Azure CLI by running the following command, which takes you to the browser to log in to Azure:

```
Az login
```

3. Run the following command to set the subscription to create the resources. If there is more than one active subscription associated with the account, run this command to set the current working subscription where all the resources will be created while executing the rest of the commands:

```
az account set --subscription <<subscription id>>
```

4. Create a resource group by running the following command. This command creates the resource group with the custom name `distelrg` in the region `westus`. We will create all our Azure resources in this dedicated Azure resource group:

```
az group create --name $resourceGroup -location
  $location
```

5. Create an Azure storage account:

```
az storage account create --location $location -name
  $storageAccount --resource-group $resourceGroup -
  kind "StorageV2" --sku "Standard_LRS"
```

6. Create a new Azure Kubernetes cluster – this can take a few minutes:

```
az aks create --resource-group $resourceGroup -name
  $clusterName --node-count 1
```

7. If you haven't already, install the Kubernetes CLI:

```
az aks install-cli
```

8. Authenticate the Kubernetes CLI:

```
az aks get-credentials --resource-group $resourceGroup
  --name $clusterName
```

9. Create an ACR account and log in to it:

```
az acr create --name $containerRegistry --resource-
  group $resourceGroup --sku Standard
```

10. Create a service principal for the container registry and register it with Kubernetes. This is an image pulling secret:

```
$acrId = $(az acr show --name $containerRegistry -
  query id --output tsv)
$acrServicePrincipalName = "$($containerRegistry)-aks-
  service-principal"
$acrSpPw = $(az ad sp create-for-rbac -name
  http://$acrServicePrincipalName --scopes $acrId -
  role acrpull --query password --output tsv)
```

```
# 6fa017c7-c249-442d-9939-0bf1a1d3051d is the app id
  of http://distelwebhostacr2-aks-service-principal
$acrSpAppId = $(az ad sp show --id 6fa017c7-c249-442d-
  9939-0bf1a1d3051d --query appId --output tsv)
$acrLoginServer = $(az acr show —name
  $containerRegistry --resource-group $resourceGroup -
  query loginServer).Trim('"')
kubectl create secret docker-registry
  $containerRegistry --namespace default --docker-
  server=$acrLoginServer --docker-username=$acrSpAppId
  --docker-password=$acrSpPw
```

11. Configure the storage account that the application is going to use by adding a new secret to Kubernetes:

```
kubectl create secret generic az-storage-acct --from-
  literal=key=$(az storage account show-connection-
  string --name $storageAccount --resource-group
  $resourceGroup --output tsv)

$acrLoginServer = $(az acr show —name
  $containerRegistry --resource-group $resourceGroup -
  -query loginServer).Trim('"')
```

12. Docker images should adhere to the ACR naming convention of `AzureContainerRegistryName.azurecr.io/NameOfTheImage:tag` for pushing images. Tag the Docker image that we created in the previous section as per the ACR naming convention using the following command:

```
docker tag distelwebhost:development
  distelwebhostacr.azurecr.io/distelwebhost:v1
```

13. Log in to the container registry that we created using the following command:

```
az acr login --name $containerRegistry
```

14. Push the Docker image to ACR using the following command:

```
docker push
  distelwebhostacr.azurecr.io/distelwebhost:v1
```

15. You can log in to the Microsoft Azure portal and search for `distelwebhostacr` and click on the resources in the results to access it from the portal, as shown in the following screenshot.

Figure 10.6 – Azure Container Registry

16. You can then click on **Repositories** and then click on the repository `distelwebhost` to see the Docker image that we pushed, as shown in the following screenshot.

Figure 10.7 – Azure Container Registry repository

In the next section, we will see how we can deploy the image from ACR to an **Azure Kubernetes cluster**.

Deploying to Azure Kubernetes

Azure Kubernetes Service offers the deployment and management of container images from ACR. In Azure Kubernetes, the smallest deployable unit that can be created and managed is called a Pod. Most Pods have one-to-one mapping with containers, but in some cases, they have one-to-many mapping with containers that share the same storage, network, and configuration specification to run the containers. To deploy the Docker image, follow these steps:

1. Create a deployment manifest YML file as shown here: `https://github.com/PacktPublishing/Distributed-.NET-with-Microsoft-Orleans/blob/main/Chapter10/Distel.WebHost/deployment.yaml`. This file will be leveraged by Azure Kubernetes for deployment and configuration. Here are the labels specified in the deployment manifest file:

    A. Labels are key-value pairs specified as attributes for objects such as Pods.

    B. Port `80` is specified to allow traffic for our Docker container.

    C. A service is specified to expose our API for public access.

    D. A load balancer is specified to allocate a public IP address that enables traffic into our Kubernetes cluster from the internet.

    E. A container and image name with tags are specified.

2. Run the following command for kubectl to point to the `distelwebhost` cluster:

```
kubectl config get-contexts
kubectl config use-context distelwebhost
```

3. Run the following command to leverage the deployment YML file and deploy the container image in an Azure Kubernetes cluster:

```
kubectl apply -f "D:\Orleans\Chapter 10\Code\
    Distel\Distel.WebHost\Deployment.yaml"
```

This command takes a few minutes as Kubernetes will pull the Docker image from ACR and instantiate.

4. Run the following command to get the external IP:

```
kubectl get svc -w
```

You will get results as shown in the following screenshot.

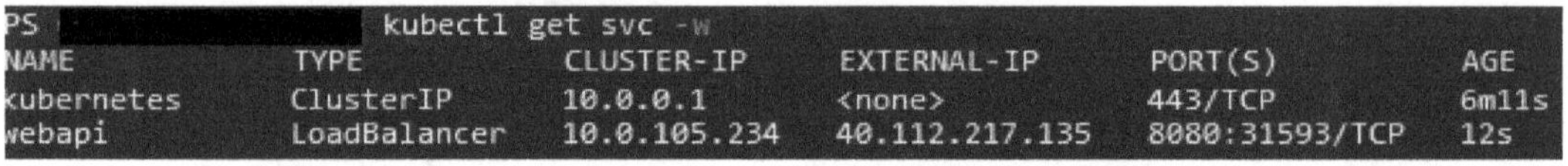

Figure 10.8 – kubectl services

5. Run the command shown in the following screenshot using an external IP to fire requests against our deployed Orleans application and verify the response:

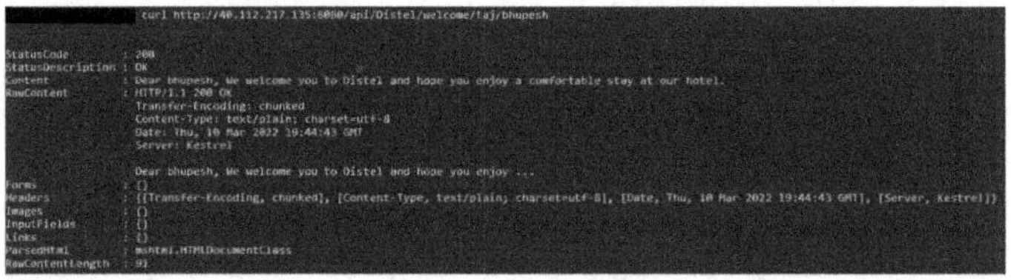

Figure 10.9 – Access the Orleans application deployed in Kubernetes

In this section, we have deployed our Orleans application in an Azure Kubernetes cluster, taking advantage of the hosting platform. You can refer to the scripts at `https://github.com/PacktPublishing/Distributed-.NET-with-Microsoft-Orleans/blob/main/Chapter10/Distel.WebHost/deployment.ps1` that were run to provision resources and for deployment. With this, we have deployed our Distel application to Azure Kubernetes using `Microsoft.Orleans.Hosting.Kubernetes`. This brings us to the end of the chapter.

Summary

In this chapter, we discussed Azure Kubernetes, leveraging the `Microsoft.Orleans.Hosting.Kubernetes` NuGet package to host an Orleans application in Kubernetes, Docker image creation, pushing a Docker image to ACR, and deploying to Azure Kubernetes. Now you have the skills required to deploy an Orleans application in Kubernetes. In the next chapter, let's learn how to deploy to Azure App Service, which is a fully managed platform as a service offering, not a container orchestrator.

Questions

1. What is the Docker platform?

    A. The Docker platform helps us to package applications (code, configuration, and dependencies) as standalone lightweight executable images.

    B. The Docker platform is an open source platform.

    C. The Docker platform supports cross-platform images and containers.

    D. All the above

    Answer – D

2. Linux and Windows are the target OSs supported by Docker.

    A. True

    B. False

    Answer – A

3. ACR is an offering to securely store and manage container images with the help of Microsoft.

    A. True

    B. False

    Answer – A

4. Which of the following statements is correct?

    A. We can push a Docker image to ACR using the `docker push` command.

    B. Azure Kubernetes Service offers deployment and management of container images from ACR.

    C. The `Microsoft.Orleans.Hosting.Kubernetes` NuGet package provides support for hosting a Microsoft Orleans application in Azure Kubernetes.

    D. All the above

    Answer – D

Further reading

We strongly recommend referring to the Orleans Kubernetes hosting documentation located here: `https://dotnet.github.io/orleans/docs/deployment/kubernetes.html`.

11
Deploying an Orleans Application to Azure App Service

In the previous chapter, we learned how to deploy an Orleans application to **Azure Kubernetes Service (AKS)**. Azure App Service is a fully managed **Platform as a Service (PaaS)** offering; it is not a container orchestrator. It provides the tools and services required to run reliable and scalable web applications that run on Microsoft Azure. App Service does not require a great deal of expertise or experience with container orchestrators or operating containers such as AKS. It provides capabilities such as autoscaling, load balancing, and security. In this chapter, we will learn how to host an Orleans silo in Azure App Service.

We will cover the following topics in this chapter:

- Introduction to Azure App Service
- Creating the required Azure resources
- Configuring the application to run on App Service
- Deploying to Azure App Service

By the end of the chapter, you will have learned all the details required to host an Orleans application in Azure App Service.

Technical requirements

To follow along, you should have the following:

- Visual Studio 2022 Community Edition

- An understanding of building .NET applications

- A basic understanding of creating Azure resources

- An Azure subscription

The code used in this chapter can be found at `https://github.com/ PacktPublishing/Distributed-.NET-with-Microsoft-Orleans/tree/ main/Chapter11`.

Introduction to Azure App Service

Azure App Service is a PaaS offering that helps to host web applications, backend services for **Single-Page Applications** (**SPAs**), and mobile applications. Applications deployed on App Service can be developed using various programming languages: C#, Java, Python, Ruby, Node.js, and PHP. It supports running applications on Windows and Linux environments. App Service brings the power of Microsoft Azure to applications, including security, autoscaling, load balancing, and automated management. App Service takes care of the patching and maintenance of the underlying environment where the application runs.

Azure App Service lets you integrate with various DevOps services, such as Azure DevOps and GitHub, to enable **continuous integration/continuous deployment** (**CI/CD**) in our application.

Coming to the pricing, you pay for the Azure compute resources you utilize with App Service. The App Service plan on which you run your apps determines the computing resources you consume.

What is included in an App Service plan?

Applications require compute resources such as servers and network components to run in Azure App Service. These compute resources are defined by Azure App Service plans. One or more applications can be configured to run on a given App Service plan. The parameters determined by an App Service plan are as follows:

- The operating system that the application runs on (Windows, Linux)
- The region where the application runs (West US, Central India, East Europe, and so on)
- The size of the **Virtual Machine** (**VM**) instances (small, medium, large)
- The number of VM instances
- The pricing tier (Free, Shared, Basic, Standard, Premium, Isolated, and so on)

  The pricing tier determines the features we get, such as custom domains, staging slots, network isolation, memory, storage, and the price we need to pay for it.

  The pricing tiers are mainly categorized as follows:

  - **Shared compute** – The application runs on a VM shared with other customers. These resources cannot scale out.
  - **Dedicated compute** – The application runs on dedicated VMs. These resources can scale out.
  - **Isolated compute** – Applications run on dedicated VMs with dedicated virtual networks. These pricing tiers provide network isolation in addition to compute isolation.

The **Azure compute unit** (**ACU**) is a notion that allows you to compare compute performance across Azure SKUs. This allows us to quickly determine which SKU is most likely to meet the performance requirements of our application.

> **Note**
>
> You can learn more about Azure compute units here: `https://docs.microsoft.com/en-us/azure/virtual-machines/acu`.

We learned in *Chapter 3, Introduction to Microsoft Orleans*, that the Orleans silo requires private ports (`siloPort` and `gatewayPort`) to communicate with other silos in the same cluster and the clients. Due to the need for private ports, Orleans applications can run only on dedicated or isolated compute. They cannot run on shared compute as they won't support private ports. In the next section, we will create all the Azure resources required to run an Orleans application on App Service.

Creating the required Azure resources

In this section, we will create all the resources required to run the Distel Orleans application in Azure App Service. The resources needed to run the Distel application are the following:

- **Azure App Service**: The compute instance that runs the silo of the Distel application
- **Azure App Service plan**: Defines the infrastructural needs of App Service
- **Azure Virtual Network**: A private network for silo-to-silo and client-to-silo communication
- **Cosmos DB**: A database to store the silo membership and the grain storage
- **Application Insights**: To capture the telemetry of the Distel application

We can create the Azure resources using any of these:

- **The Azure CLI (Command-Line Interface)**: A cross-platform CLI to connect to Azure and administer Azure resources.
- **Azure Resource Manager (ARM)**: Provides a management layer to manage Azure resources. The details of resources to be managed are passed in declarative JSON syntax called ARM templates.
- **Bicep**: A domain-specific language that simplifies resource authoring with type safety and cleaner syntax. It is a transparent abstraction over ARM templates.
- **The Azure portal**: We can also create and manage Azure resources from the Azure portal, but this is not recommended to create production resources.

Creating the resources using the Azure CLI, ARM, and Bicep can be integrated with continuous integration pipelines. They all have feature parity.

To create the resources for the Distel application, let's use the Azure CLI:

1. First, we need to install the Azure CLI. To install the Azure CLI, download the Azure CLI installer from `https://docs.microsoft.com/en-us/cli/azure/install-azure-cli`.

2. Once the installation is complete, open the **PowerShell** window and follow the following steps to create the required resources.

3. Log in to the Azure CLI by running the following command; this takes you to the browser to log in to Azure:

   ```
   Az login
   ```

4. Run the following command to set the subscription to create the resources:

   ```
   az account set --subscription <<subscription id>>
   ```

   If there is more than one active subscription associated with the account, run this command to set the current working subscription where all the resources will be created while executing the rest of the commands.

5. Create a resource group by running the following command:

   ```
   az group create --location westus --name rg-distel-prod
   ```

   This command creates the resource group with the name `rg-distel-prod` in the `westus` region:

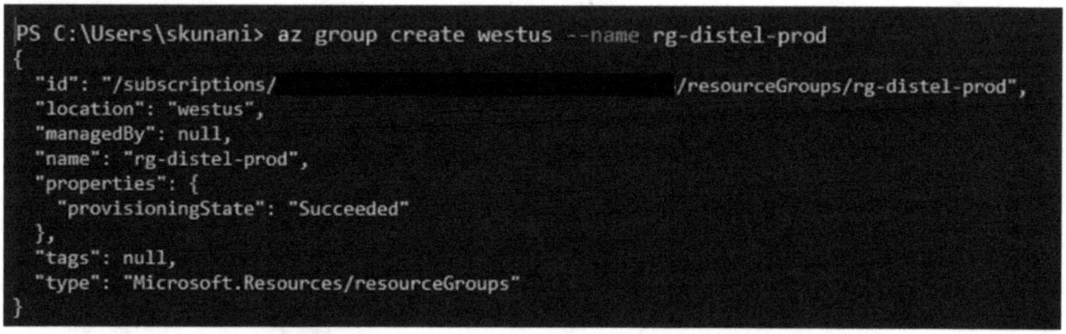

```
PS C:\Users\skunani> az group create westus --name rg-distel-prod
{
  "id": "/subscriptions/                                    /resourceGroups/rg-distel-prod",
  "location": "westus",
  "managedBy": null,
  "name": "rg-distel-prod",
  "properties": {
    "provisioningState": "Succeeded"
  },
  "tags": null,
  "type": "Microsoft.Resources/resourceGroups"
}
```

Figure 11.1 – Created Azure resource group

6.   Let's create the App Service plan by running the following command:

```
az appservice plan create --resource-group rg-distel-
prod  --name appplan-distel --number-of-workers 4 -
sku S1
```

By running this command, we will create an App Service plan with the name
appplan-distel using the SKU S1, which is from the dedicated compute
category, and with four nodes in the clusters.

7.   Now create the virtual network for the communication:

```
az network vnet create --resource-group rg-distel-prod
--name vnet-prod-westus --address-prefix 10.0.0.0/16
--subnet-name sent-prod-westus --subnet-prefix
10.0.0.0/24
```

This private network is required for all the silos' communication. In this command,
we are creating a virtual network with address space 10.0.0.0 /16 and defining
the subnet with address space 10.0.0.0/24.

> **Note**
>
> Azure **Virtual Network** (**VNet**) is the fundamental component of your
> Azure private network. It enables many different Azure resources to securely
> connect with one another, the internet, and on-premises networks. Refer to the
> Azure documentation to learn more about Azure VNet: https://docs.
> microsoft.com/en-us/azure/virtual-network/virtual-
> networks-overview.

8.   Now it is time to create the Azure web app by running the following command:

```
az webapp create --resource-group rg-distel-prod -
name app-distel --plan appplan-distel --vnet vnet-
prod-westus --subnet sent-prod-westus
```

Here, we are creating the Azure web app with the Azure App Service plan we
created in *Step 4*. We are also associating the virtual network we created in *Step 5*.

9. Configure the web app created in previous step to have private ports for `siloPort` and `gatewayPort` for silo-to-silo and client-to-silo communication respectively:

```
az webapp config set --resource-group rg-distel-prod -
-name app-distel --generic-configurations
"{\"vnetPrivatePortsCount\": 2}"
```

Run the preceding command to enable private ports. Here, we are configuring to have only two ports, one for `siloPort` and the second one for `gatewayPort`.

10. Create a Cosmos DB resource, which will be used as storage for cluster management and grain storage:

```
az cosmosdb create --resource-group rg-distel-prod -
name cosmos-distel-prod
```

In this command, we are creating the Azure Cosmos DB resource with the name `cosmos-distel-prod` in the `rg-distel-prod` resource group.

11. *Create Application Insights resources*: An Application Insights resource is required to capture the telemetry logs. Create an Application Insights resource by running the following command:

```
az monitor app-insights component create --app ai-
distel-prod --location westus  --resource-group rg-
distel-prod
```

This command will create the Application Insights resource with the name `ai-distel-prod`.

12. Now get the instrumentation key and set that as an `environment` variable to Azure App Service so that the Distel application can use that to configure Application Insights. Run the following command to fetch the instrumentation key:

```
[String] $instrumentationKey = (az monitor app-insights
component show --app ai-distel-prod --resource-group
rg-distel-prod --query  "instrumentationKey" --output
tsv)
```

13. Once the instrumentation key is fetched, run the following command to set the environment variable:

```
az webapp config appsettings set --resource-group rg-
distel-prod --name app-distel --settings
APPINSIGHTS_INSTRUMENTATIONKEY=$instrumentationKey
```

This command will set the APPINSIGHTS_INSTRUMENTATIONKEY environment variable with the instrumentation key of ai-distel-prod.

After creating all the resources, do the following:

1. Go to the Azure portal and select the **rg-distel-prod** resource group. This will show all the resources created under this resource group.

2. Now select **Export Template**. This will generate the ARM template with parameters that you can use to recreate the same resources. You can also customize this template as per your needs and use it in CI/CD pipelines to create resources.

3. On the same **Export Template** page, if you select **Visualize template**, you can see resources created in this section as shown in *Figure 11.2*:

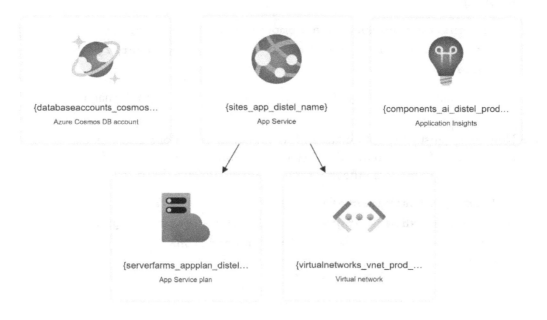

Figure 11.2 – Azure resource topology of Distel

> **Note**
>
> The Azure CLI command reference list can be found here: `https://docs.microsoft.com/en-us/cli/azure/service-page/list%20a%20-%20z?view=azure-cli-latest`.

In this section, we have created all the resources required to run our Distel application on Azure App Service with four silos. In the next section, let's update our Distel application to run on the Azure resources created in this section.

Configuring the application to run on App Service

The Distel application we have built has the silo configuration to run on the local developer's machine. It uses the in-memory storage for cluster membership storage and the silo endpoint is configured as `IPAddress.Loopback`. This configuration is only used on a local developer machine. This configuration will not work in production scenarios where we have a cluster of silos as they all need to communicate with each other. The grain state should also be persisted in permanent storage rather than being stored in temporary in-memory storage since we'll lose all the state if the node goes down.

To make the Distel application run in Azure App Service, we need to update the silo endpoint configuration. We also need to configure the grain storage to use Cosmos DB. The following steps will guide you to make our application ready to deploy to Azure App Service:

1. *Configure the silo endpoint*: Let's start by configuring the silo endpoints to use the local private IP address and the private ports.

   The private IP address and the private ports are available to the application through environment variables in Azure App Service. Those environment variables are as follows:

   - `WEBSITE_PRIVATE_IP`: This environment variable holds the value of the private IP address. The endpoint IP address can be retrieved from the environment variable as shown in the following code snippet:

   ```
   var privateIP =
     Environment.GetEnvironmentVariable
     ("WEBSITE_PRIVATE_PORTS")
   ```

- WEBSITE_PRIVATE_PORTS: This environment variable holds the private ports separated by a comma. We can extract ports used to configure silos and gateways using the following code snippet and configure the silo endpoint using the private IP and ports retrieved:

```
var strPorts =
   Environment.GetEnvironmentVariable
   ("WEBSITE_PRIVATE_PORTS").Split(',');
if (strPorts.Length >= 2)
{
    siloPort = int.Parse(strPorts[0]);
    gatewayPort = int.Parse(strPorts[1]);
}
siloBuilder.ConfigureEndpoints(IPAddress.Parse(private
   IP), siloPort: siloPort, gatewayPort: gatewayPort);
```

> **Note**
>
> The list of default environment variables available in App Service can be found here: https://docs.microsoft.com/en-us/azure/app-service/reference-app-settings.

2. *Configure the membership storage*: We learned in *Chapter 4, Understanding Grains and Silos*, that Orleans supports various providers to store the cluster membership. In this chapter, we will use the Cosmos DB provider to store the cluster membership. Use the following code snippet to configure the cluster Cosmos DB membership:

```
siloBuilder.UseCosmosDBMembership(options =>
{
    options.AccountEndpoint = "<<Cosmos DB
       endpoint>>";
    options.AccountKey = "Cosmos DB account key";
    options.Collection = "Membership";
    options.CanCreateResources = true;
});
```

In the preceding code, we are configuring the Cosmos DB endpoint and the account key. The name of the Cosmos DB collection we use for membership storage is set as Membership. By setting the CanCreateResources property to true, we are instructing the Orleans runtime to create resources if they do not exist. That means Orleans will create the Membership collection if it is not present while running the application.

3. *Configure the grain storage*: We learned about the various storage providers and about creating a custom storage provider in *Chapter 5, Persistence in Grains*. Here, we will use a Cosmos DB grain storage provider. Configure the grain storage using the Cosmos DB provider by using the following code snippet:

```
siloBuilder.AddCosmosDBGrainStorageAsDefault
    (options =>
    {
        options.AccountEndpoint = "<<Cosmos DB
            endpoint>>";
        options.AccountKey = "Cosmos DB account key";
        options.Collection = "State";
        options.CanCreateResources = true;
    });
```

Like the cluster Membership storage provider we have configured the Cosmos DB provider with the endpoint and account key. We have named the grain storage collection State.

We are using the same Cosmos DB resource to store both cluster membership and grain storage. But it is often seen as best practice to have separate instances following the **Separation of Concerns** design principle.

4. Now configure the Distel application to fetch the Application Insights key from environment variables as shown in the following code snippet:

```
builder.Services.AddApplicationInsightsTelemetry
    (Environment.GetEnvironmentVariable
    ("APPINSIGHTS_INSTRUMENTATIONKEY"));
builder.Host.UseOrleans(siloBuilder =>
    {
    // Code removed for brevity
        siloBuilder
        .ConfigureLogging(logging =>
```

```
        logging.AddApplicationInsights
        (Environment.GetEnvironmentVariable
        ("APPINSIGHTS_INSTRUMENTATIONKEY")));
    });
```

> **Note**
>
> To configure the cluster membership and storage, we have the Cosmos DB connection secret configured in plain text in the code. This is done only for demonstration purposes. In production scenarios, we should be fetching such secrets from secret providers such as Azure Key Vault.

The Distel application also needs a storage provider for `PubSubStore`, which is required for Orleans streams. We will leave it as an exercise for you to configure the storage for `PubSubStore`.

Geographically distributing the application

For application scenarios that demand high scale and users spread across multiple regions, we should have the application deployed in multiple regions. Here, we have created the resources required for the Distel application in only one region. To make this geographically distributed, we will have the Distel application running in multiple regions by creating a similar infrastructure in multiple regions. We may choose the regions to run the application by considering various factors, such as *user load* and *availability zones*. We will have a traffic manager configured to point at individual app services. The traffic manager will automatically handle distributing the user requests across all the regions.

> **Note**
>
> More about running geo-distributed Azure App Service can be found here: `https://docs.microsoft.com/en-us/azure/app-service/environment/app-service-app-service-environment-geo-distributed-scale.`

In the case of multi-region applications, we may have region-specific storage resources. In this scenario, we need to choose the grain effectively because grains with the same identity can be activated in multiple regions simultaneously. We need to consider the right consistency strategy based on the storage provider we use. In Cosmos DB, the consistency levels we have are *Strong, Bounded staleness, Session, Consistent Prefix*, and *Eventual*. Refer to the documentation (`https://docs.microsoft.com/en-us/azure/cosmos-db/consistency-levels`) to choose the right consistency level for your scenario.

In multi-region deployments, the name of the region is `ClusterId`:

```
siloBuilder.Configure<ClusterOptions>(opts =>
{
    opts.ClusterId =
       Environment.GetEnvironmentVariable("REGION_NAME");
    opts.ServiceId = "DistelAPI";
});
```

In the preceding code snippet, the region name is fetched from the App Service `REGION_NAME` environment variable and set as `ClusterId` in `ClusterOptions` while building the silo.

In this section, we have configured the Distel application to run on the Azure App Service infrastructure we created in the *Creating the required Azure resources* section. In the next section, let's publish and run the Distel application from Azure App Service.

Deploying an Orleans application to App Service

Up until now, we have learned how to build and test an Orleans application. In the previous section of this chapter, we configured the application to be deployed to Azure App Service. Now it is time to deploy the application and see it in action. We will be deploying our Distel application using the CLI from a local machine. This is done purely for demonstration purposes. For production scenarios, we use DevOps pipelines such as Azure DevOps and Jenkins. Follow these steps to package and deploy the Distel application:

1. Run the `dotnet publish` command from the Distel solution root path to generate the publish artifacts as shown here:

    ```
    dotnet publish --configuration RELEASE
    ```

    This command will build and generate the artifacts for the `RELEASE` configuration. The generated artifacts will be published to the `~\Distel.WebHost\bin\Release\net6.0\publish` folder.

2. Now we will package the published artifacts to a ZIP file with the name `Distel.Zip` using the following PowerShell command:

    ```
    Compress-Archive -Path
    .\Distel.WebHost\bin\Release\net6.0\publish\ * -
    DestinationPath Distel.zip
    ```

3. It is now time to publish the `Distel.Zip` package to Azure App Service by running the following command:

    ```
    az webapp deploy --resource-group rg-distel-prod -
    name app-distel  --src-path Distel.zip
    ```

    This command will publish the package we created, `Distel.zip`, to Azure App Service.

4. Once it is done publishing, we can browse the Swagger UI of the Distel application by navigating to the App Service URL `https://<<app-service-name>>.azurewebsites.net/swagger/index.html`. This will show the same Swagger documentation UI as we saw in the previous chapter.

5. From here, we can execute the APIs exposed by the Distel application. If we execute the welcome action `/api/Distel/welcome/{hotel}/{guestname}`, we will see the output as shown in the following screenshot:

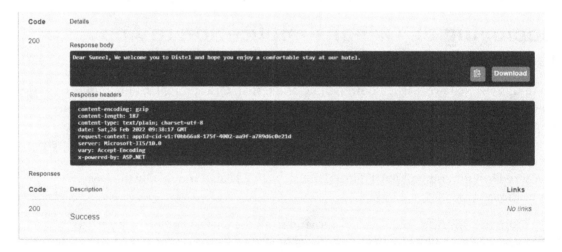

Figure 11.3 – Output of the welcome API

6. Let's also make a call to an API action that persists state, for example, `/api/Distel /checkin/{hotel}` (for reference, check *Chapter 5, Persistence in Grains*).

Now when we look at the data explorer of the Cosmos DB resource we created, we will see two collections created, named `Membership` and `State`, as configured in the silo configuration. They persist the details of silo membership and grain state respectively.

The `Membership` collection will have cluster membership entries for each silo as shown in *Figure 11.4*. Since we have four nodes in the cluster, there are four entries in the `Membership` collection. Each entry for the silo stores its IP address, silo port number, gateway port number, and the cluster ID.

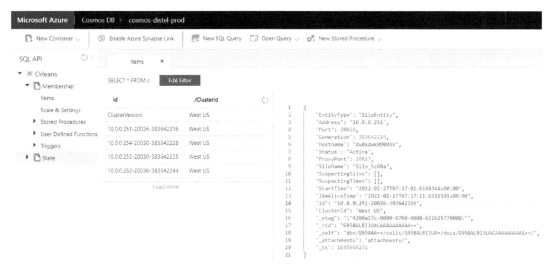

Figure 11.4 – Silo membership details

Let's now look at the `State` collection. Each item in this collection corresponds to the state of a grain type. In *Figure 11.5*, we can see the `checkedInGuests` state of `HotelGrain`. Along with the state, it contains the grain type, which contains the state name of the grain.

Figure 11.5 – Persisted grain state

Let's now navigate to the Application Insights resource to see the telemetry captured by the Distel application. When you go to **Application Insights Transaction** search, you will see the telemetry of the requests to the ASP.NET core application and dependencies, along with the telemetry coming from the Orleans runtime, as shown in *Figure 11.6*:

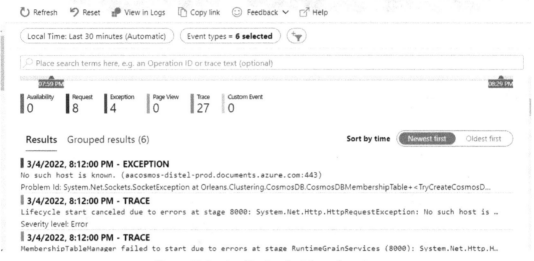

Figure 11.6 – Application Insights telemetry

The **EXCEPTION** we see in *Figure 11.6* is due to the misconfiguration in the Membership Cosmos DB configuration for the demonstration.

To summarize, in this section, we deployed our Distel application to the App Service infrastructure, then saw the application in action by executing the APIs exposed through Swagger UI. Next, we saw how Orleans maintains the membership table of the silos part of the cluster and persisted grain state.

Summary

In this chapter, we learned how to create the resources required to run the Orleans application on Azure App Service in the Azure CLI. Next, we modified the silo configuration of the Distel application we built in the previous chapter to run on Azure App Service. We deployed the application to the App Service infrastructure and saw it in action.

We hope this book has helped you to gain the required skills to build highly scalable distributed applications using Microsoft Orleans. There are further topics that you can explore by referring to the notes and the *Further reading* sections of the chapters. With Distel, we have covered a few scenarios to demonstrate the various concepts of Orleans, and it can be further extended to a full-fledged hotel management application leveraging the potential of Orleans.

We wish you the best for your next distributed application using Microsoft Orleans.

Happy learning!!!

Questions

1. Can an Orleans application run on the Free Tier of Azure App Service?

    A. Yes

    B. No

    Answer – B

2. What is the environment variable that holds the private ports in Azure App Service?

    A. WEBSITE_PORTS

    B. WEBSITE_PRIVATE_PORTS

    C. PORTS

    D. PRIVATE_PORTS

    Answer – B

3. We can use UseLocalhostClustering to configure a silo for both development and production scenarios.

    A. TRUE

    B. FALSE

    Answer – B

4. The number of private ports required to configure an Orleans silo is…

    A. 1

    B. 2

    C. 3

    D. 4

    Answer – B

5. The telemetry from the Orleans runtime can be pushed to Application Insights.

    A. TRUE

    B. FALSE

    Answer – A

Index

`Packt.com`

Subscribe to our online digital library for full access to over 7,000 books and videos, as well as industry leading tools to help you plan your personal development and advance your career. For more information, please visit our website.

Why subscribe?

- Spend less time learning and more time coding with practical eBooks and Videos from over 4,000 industry professionals

- Improve your learning with Skill Plans built especially for you

- Get a free eBook or video every month

- Fully searchable for easy access to vital information

- Copy and paste, print, and bookmark content

Did you know that Packt offers eBook versions of every book published, with PDF and ePub files available? You can upgrade to the eBook version at `packt.com` and as a print book customer, you are entitled to a discount on the eBook copy. Get in touch with us at `customercare@packtpub.com` for more details.

At `www.packt.com`, you can also read a collection of free technical articles, sign up for a range of free newsletters, and receive exclusive discounts and offers on Packt books and eBooks.

Other Books You May Enjoy

If you enjoyed this book, you may be interested in these other books by Packt:

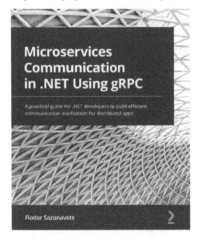

Microservices Communication in .NET Using gRPC

Fiodar Sazanavets

ISBN: 978-1-80323-643-8

- Get to grips with the fundamentals of gRPC and Protobuf
- Debug gRPC components inside a .NET application to locate and fix errors
- Understand gRPC best practices, such as performance enhancement
- Effectively translate between gRPC and native C# code by applying well-known types
- Secure gRPC communication inside a .NET application
- Discover how to monitor gRPC on .NET by applying logging and metrics

Enterprise Application Development with C# 9 and .NET 5

Ravindra Akella, Arun Kumar Tamirisa, Suneel Kumar Kunani,
Bhupesh Guptha Muthiyalu

ISBN: 978-1-80020-944-2

- Design enterprise apps by making the most of the latest features of .NET 5

- Discover different layers of an app, such as the data layer, API layer, and web layer

- Explore end-to-end architecture, implement an enterprise web app using .NET and C# 9, and deploy the app on Azure

- Focus on the core concepts of web application development such as dependency injection, caching, logging, configuration, and authentication, and implement them in .NET 5

- Integrate the new .NET 5 health and performance check APIs with your app

- Understand how .NET 5 works and contribute to the .NET 5 platform

Packt is searching for authors like you

If you're interested in becoming an author for Packt, please visit `authors.packtpub.com` and apply today. We have worked with thousands of developers and tech professionals, just like you, to help them share their insight with the global tech community. You can make a general application, apply for a specific hot topic that we are recruiting an author for, or submit your own idea.

Hi!

We're Bhupesh and Suneel, the authors of *Distributed .NET with Microsoft Orleans*. We really hope you enjoyed reading this book and found it useful for increasing your productivity and efficiency in building distributed applications.

It would really help us (and other potential readers!) if you could leave a review on Amazon sharing your thoughts on *Distributed .NET with Microsoft Orleans*.

Go to the link below or scan the QR code to leave your review:

`https://packt.link/r/1801818975`

Your review will help us to understand what's worked well in this book, and what could be improved upon for future editions, so it really is appreciated.

Best Wishes,

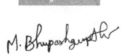